Application of Materials Science in the Study of Cultural Heritage

Application of Materials Science in the Study of Cultural Heritage

Editors

Marco Martini
Anna Galli

MDPI • Basel • Beijing • Wuhan • Barcelona • Belgrade • Manchester • Tokyo • Cluj • Tianjin

Editors
Marco Martini
Università degli Studi
Milano-Bicocca
Italy

Anna Galli
Università degli Studi
Milano-Bicocca
Italy

Editorial Office
MDPI
St. Alban-Anlage 66
4052 Basel, Switzerland

This is a reprint of articles from the Special Issue published online in the open access journal *Applied Sciences* (ISSN 2076-3417) (available at: https://www.mdpi.com/journal/applsci/special_issues/Materials_Science_Cultural_Heritage).

For citation purposes, cite each article independently as indicated on the article page online and as indicated below:

LastName, A.A.; LastName, B.B.; LastName, C.C. Article Title. *Journal Name* **Year**, *Volume Number*, Page Range.

ISBN 978-3-0365-4561-5 (Hbk)
ISBN 978-3-0365-4562-2 (PDF)

© 2022 by the authors. Articles in this book are Open Access and distributed under the Creative Commons Attribution (CC BY) license, which allows users to download, copy and build upon published articles, as long as the author and publisher are properly credited, which ensures maximum dissemination and a wider impact of our publications.

The book as a whole is distributed by MDPI under the terms and conditions of the Creative Commons license CC BY-NC-ND.

Contents

About the Editors vii

Marco Martini and Anna Galli
Special Issue: Application of Materials Science in the Study of Cultural Heritage
Reprinted from: *Appl. Sci.* **2022**, *12*, 5069, doi:10.3390/app12105069 1

Josefina Pérez-Arantegui and Paz Marzo
Characterization of Islamic Ceramic Production Techniques in Northeast Iberian Peninsula: The Case of Medieval Albarracín (Spain)
Reprinted from: *Appl. Sci.* **2021**, *11*, 7212, doi:10.3390/app11167212 5

Ivana Angelini, Cinzia Bettineschi, Marica Venturino and Gilberto Artioli
Gaming in Pre-Roman Italy: Characterization of Early Ligurian and Etruscan Small Pieces, Including Dice
Reprinted from: *Appl. Sci.* **2022**, *12*, 2130, doi:10.3390/app12042130 21

Sara Fiorentino, Tania Chinni and Mariangela Vandini
Materials Inspiring Methodology: Reflecting on the Potential of Transdisciplinary Approaches to the Study of Archaeological Glass
Reprinted from: *Appl. Sci.* **2021**, *11*, 8049, doi:10.3390/app11178049 37

Robert H. Tykot
Non-Destructive pXRF on Prehistoric Obsidian Artifacts from the Central Mediterranean
Reprinted from: *Appl. Sci.* **2021**, *11*, 7459, doi:10.3390/app11167459 51

Marco Martini and Anna Galli
Thermoluminescence Analysis of the Clay Core of Bronze Statues: A Re-Appraisal of the Case Studies of Lupa Capitolina and Other Masterpieces in Rome
Reprinted from: *Appl. Sci.* **2021**, *11*, 7820, doi:10.3390/app11177820 71

Alice Dal Fovo, Jana Striova, Enrico Pampaloni and Raffaella Fontana
Unveiling the Invisible in Uffizi Gallery's Drawing 8P by Leonardo with Non-Invasive Optical Techniques
Reprinted from: *Appl. Sci.* **2021**, *11*, 7995, doi:10.3390/app11177995 85

About the Editors

Marco Martini

Prof. Marco Martini, Full Professor of Applied Physics since 2002 at the Department of Materials Science, University of Milano-Bicocca. Director of the same department from 2012 to 2018, and Director of the Interdepartmental Center, Bi-PAC for the Study and Protection of Historical Artistic Cultural Heritage since 2017. His scientific interests concern the study of defects in insulating materials and their optical properties, in particular luminescence and the dosimetry of ionizing radiation. Closely derived from the skills acquired in the aforementioned studies is the interest in physics applied to cultural heritage.

Anna Galli

Prof. Anna Galli, Associate Professor of Applied Physics at the Department of Materials Science, University of Milano-Bicocca. Graduated in physics and specialized in materials science; her research topics focus on radiation physics linked to dosimetry and spectroscopy applied to cultural heritage. Scientific coordinator of LAMBDA (Laboratory of Milano Bicocca University for Dating and Archeometry) at the Department of Materials Science of University of Milano-Bicocca. Since 2020 until present, Member of the Scientific Board of Associazione Italiana di Archeometria (AIAr).

Editorial

Special Issue: Application of Materials Science in the Study of Cultural Heritage

Marco Martini and Anna Galli *

Dipartimento di Scienza dei Materiali, Università degli Studi di Milano-Bicocca, Via R. Cozzi 55, 20125 Milano, Italy; m.martini@unimib.it
* Correspondence: anna.galli@unimib.it

1. Introduction

The application of advanced techniques to the study of ancient materials has been increasingly demonstrating to be fundamental to a deeper knowledge of artistic and historic artefacts, contributing to their conservation and restoration. The number of scientific methodologies applied to the Cultural Heritage is huge and difficult to be listed. However, it can be noted that an important role is played by Materials Science: scientific techniques developed in Materials Science allow to contribute a multidisciplinary approach in Archaeology, History of Art, and Conservation Science. By studying the materials that constitute the artefact, deeper information can be reached, relatively to the work of art, such as the elements and compounds by which it was made and their level of degradation over time. The final goal is the possibility of determining the chronology of the making of the various part of the work of art, its provenance, the techniques of realization, the attribution to an author, and the way of intervention for restoration. Materials Science offers many different scientific methodologies in order to investigate ancient materials and artefacts. The current Special Issue collects papers dealing with the applications of Materials Science to the different types of human artefacts such as ceramics, glass, paintings, metal objects. This Special Issue includes six papers that were accepted through a stringent reviewing process and they are summarized in the following section. The contributions topics range from instrumentation and technical developments to case studies, and to methodological innovations. The authors have exploited many characteristic analytical methods of Materials Science: imaging techniques (IRR, RX, tomography); traditional as well as innovative dating techniques like various luminescence methods; spectroscopic techniques (PIXE, Raman, FTIR, UV-vis reflectance spectroscopy), and their synergic association.

2. Review of Issue Contents

Artefacts studies are concerned with the investigation of the overall life cycle from the acquisition and processing of the raw materials through production and distribution, to use, reuse, and discard. Thus, Material Science is involved in the reconstruction of how the artefacts were made, where they were made and distributed, and how they were used.

For ceramics, in this Special Issue, Pèrez-Arantegui et al. [1] have contributed a study of Islamic ceramic manufacture in northern al-Andalus (the Muslim part of the Iberian Peninsula) with the aim to improve general knowledge of small production centres and add new insights into medieval ceramic technology in the Peninsula. This work characterised nearly one hundred ceramic fragments, dated since the 11th century CE from archaeological sites in Albarracin, all of them decorated, although only the results of glazed pottery are discussed. Most of them belonged to the tableware group (bowls, beakers, and dishes); the ceramic bodies were analysed by Inductively Coupled Plasma-Optical Emission Spectrometry (ICP-OES), whereas the glazes were examined with Optical Microscopy (OM), Scanning Electron Microscopy (SEM), and Inductively Coupled Plasma-Quadrupole Mass Spectrometer (ICP-QMS). The obtained results allowed the characterisation of the

Citation: Martini, M.; Galli, A. Special Issue: Application of Materials Science in the Study of Cultural Heritage. *Appl. Sci.* **2022**, *12*, 5069. https://doi.org/10.3390/app12105069

Received: 11 May 2022
Accepted: 16 May 2022
Published: 18 May 2022

Publisher's Note: MDPI stays neutral with regard to jurisdictional claims in published maps and institutional affiliations.

Copyright: © 2022 by the authors. Licensee MDPI, Basel, Switzerland. This article is an open access article distributed under the terms and conditions of the Creative Commons Attribution (CC BY) license (https://creativecommons.org/licenses/by/4.0/).

pottery production from Albarracin Taifa in the 11–12th centuries CE, and established differences and similarities with Islamic productions manufactured in other Taifa kingdoms of the Iberian Peninsula during the same period. The analyses carried out on the ceramic bodies confirmed that the selection of raw materials depended on the decoration to be applied afterward.

Always remaining in the ceramics world, but moving in time and in space, Angelini et al. [2] reported the study of very unusual ceramic dice found in Castello d'Annone (Alessandria, Italy), bearing Etruscan letters and randomly distributed dots. They are a *unicum* and confirmed their exceptional importance both for the large quantity of recovered finds and for the reconstruction of the social and cultural context of an "emporium". Based on the measured data and the archaeological information, it was proposed that the ceramic pieces, in the form of washers, spheres, and uninscribed cubes, were used as counters in board games.

The contribution that Materials Science has made to the understanding of ancient glassmaking is unquestionable, as research undertaken in recent decades has extensively demonstrated. Archaeological glass is far from being a homogeneous class of materials, encompassing objects made for different uses, manufactured in different periods and geographic areas, and with a variety of tools and working techniques. If all these factors are not adequately considered when approaching the study of ancient glasses from an archaeometric perspective, data obtained by analyses can incur the risk of being less informative or even misinterpreted. Moving from previously performed research, the paper by Fiorentino et al. [3] is focused on the potential of synergistic approaches for the study of archaeological glasses, based on the interrelation among different disciplines and fostering the integration of archaeological and historical knowledge with data-driven scientific analyses.

The obsidians are a very particular subset of archaeological glasses: they are igneous glassy rocks, formed during volcanic eruptions during the past 20 million years. While elemental analyses were used for identifying specific sources starting in the 1960s, the development of non-destructive and especially portable X-ray fluorescence spectrometers has revolutionized the number of artifacts tested since 2010, allowing statistically significant numbers for potential comparisons based on variables including time period, open-water distance, visual and physical properties, and cultural contexts. Tykot in his paper [4] illustrates the importance of this kind of study, highlighting one overall accomplishment: the documentation of long-distance travel routes, based on the distribution proportions and quantity of obsidian artefacts from the different geological sources.

The ability of luminescence techniques (TL and OSL) to provide dating if applied to ceramics or sediments is known worldwide, but not everyone knows that in some cases it is possible to obtain an indirect dating of the bronze statues by dating the clay core. The basic idea of dating by TL can be applied, in principle, to the material remaining in the interior of a bronze statue after its casting, the so-called clay core. In favourable cases, this material behaves like a ceramic, and the procedures used for dating ceramics can be also applied to clay cores. This is extremely important, considering that, with very few exceptions that are not treated in the Martini and Galli paper [5], metal objects cannot be dated by absolute techniques. In [5], the application of TL dating to clay cores was introduced and the specific difficulties deriving from the characteristics of this material, together with the complex determination of the radiation environment, are commented on.

The systematic application of complementary analytical methods typical of Material Science proved essential for a deeper understanding of such complex and extremely valuable artworks. The paper of Dal Fovo et al. [6] is an example of how the optical techniques (e.g., multispectral reflectography in the visible (Vis) and near-infrared (NIR) and optical coherence tomography), due to their non-invasive characteristic, are the best practice to approach to extremely fragile drawings on the paper. The authors report on the application of non-invasive multi-modal optical analysis on a double-side drawn sheet by Leonardo,

Drawing 8P from the Uffizi Gallery of Florence, Italy, offering a useful contribution to the literature on the drawing technique used by Leonardo in his early production.

3. Conclusions

This Special Issue highlights how Materials Science usefully participates in the study of Cultural Heritage allowing researchers to deeply understand the history of works of art, and how material processing influences their structure, properties, and performances.

Funding: This research received no external funding.

Conflicts of Interest: The authors declare no conflict of interest.

References

1. Pérez-Arantegui, J.; Marzo, P. Characterization of Islamic Ceramic Production Techniques in Northeast Iberian Peninsula: The Case of Medieval Albarracin (Spain). *Appl. Sci.* **2021**, *11*, 7212. [CrossRef]
2. Angelini, I.; Bettineschi, C.; Venturino, M.; Artioli, G. Gaming in Pre-Roman Italy: Characterization of Early Ligurian and Etruscan Small Pieces, Including Dice. *Appl. Sci.* **2022**, *12*, 2130. [CrossRef]
3. Fiorentino, S.; Chinni, T.; Vandini, M. Materials Inspiring Methodology: Reflecting on the Potential of Transdisciplinary Approaches to the Study of Archaeological Glass. *Appl. Sci.* **2021**, *11*, 8049. [CrossRef]
4. Tykot, R. Non-Destructive pXRF on Prehistoric Obsidian Artifacts from the Central Mediterranean. *Appl. Sci.* **2021**, *11*, 7459. [CrossRef]
5. Martini, M.; Galli, A. Thermoluminescence Analysis of the Clay Core of Bronze Statues: A Re-Appraisal of the Case Studies of Lupa Capitolina and Other Masterpieces in Rome. *Appl. Sci.* **2021**, *11*, 7820. [CrossRef]
6. Dal Fovo, A.; Striova, J.; Pampaloni, E.; Fontana, R. Unveiling the Invisible in Uffizi Gallery's Drawing 8P by Leonardo with Non-Invasive Optical Techniques. *Appl. Sci.* **2021**, *11*, 7995. [CrossRef]

Article

Characterization of Islamic Ceramic Production Techniques in Northeast Iberian Peninsula: The Case of Medieval Albarracin (Spain)

Josefina Pérez-Arantegui [1,*] and Paz Marzo [2]

1. Instituto Universitario de Investigación en Ciencias Ambientales de Aragón (IUCA), Universidad de Zaragoza, 50009 Zaragoza, Spain
2. Departamento Química Analítica, Facultad de Ciencias, Universidad de Zaragoza, 50009 Zaragoza, Spain; pazmarzo@yahoo.es
* Correspondence: jparante@unizar.es

Abstract: Ceramic objects found in the Islamic *Taifa* of Albarracin (Spain), 12th century CE, were studied to ascertain the main characteristics and influences of its manufacture. Production centers even from small kingdoms can add new insights in medieval ceramic technology. Several types of decorated ceramics, such as tin-opacified glazed, monochrome glazed and *cuerda seca*, were investigated. Ceramic bodies were analyzed by ICP-Optical Emission Spectrometry, and glazes were studied by Scanning Electron Microscopy with Energy Dispersive X-ray Spectrometry. All the ceramic bodies showed the use of Ca-rich pastes, although three groups could be distinguished and related to their decoration. Lead and silicon were the main components of the glazes, as well as scattered tin oxide in the case of white or green opacified glazes. Some features, such as calcareous bodies, double firing for tin-opacified glazes, glaze components, and coloring oxides, were common in Albarracin samples and other Islamic production centers in the Iberian Peninsula. However, some differences were also highlighted in lead/silica proportions and *cuerda seca* decoration, and several influences from northern or southern pottery centers. Lead isotope ratios, measured by ICP-Quadrupole Mass Spectrometry, revealed two different sources or suppliers of lead raw materials according to the type of glaze to be prepared.

Keywords: characterization; ceramic; glaze; tin opacified; *cuerda seca*; lead isotopes; SEM; ICP-QMS; Islamic

1. Introduction

Islamic ceramic manufacture in northern al-Andalus (the Muslim part of the Iberian Peninsula) is less known; however, the study of production centers from some small, but splendorous, kingdoms in the Peninsula can improve general knowledge and add new insights into medieval ceramic technology. The Iberian Peninsula was part of the Islamic culture since the 8th century CE, but it was in the 10th century CE when it reached its political independence from the Damascus caliphate. At the beginning of the 11th century CE, al-Andalus was divided in small kingdoms (called *Taifas*), such as Toledo, Cordoba, Granada, or Zaragoza (Figure 1). This was the case of Albarracin *Taifa*. Nowadays, Albarracin is a town located in the NE of the Iberian Peninsula, in a mountain range in the current Aragon. Its good situation in an environment with lots of natural resources (water, vegetation and animals) has favored human habitation since ancient periods, as is shown by the rock art found in the surroundings that dates from the 6th millennium BCE [1]. Albarracin *Taifa* was governed by a Berber family, the Banu-Razin, and it had a flourishing cultural and economic development which reflected the influence of the Muslim society in different aspects of its way of life and also in ceramic production [2,3]. At the beginning of the 12th century CE, the kingdom was under the influence of the Almoravid

political power from Valencia and Murcia. At the end of the 12th century CE, in 1170, Albarracin was given to the Christian Azagra family, but at the same time it preserved its independence from Aragon and Castile during two centuries; this allowed it to continue economic relations not only with northern Iberian regions dominated by Christians but also with Muslim kingdoms [2,3]. All these features reveal a great interest on the study of ceramic productions in Albarracin *Taifa* in order to understand pottery characteristics, technology and actual exchanges between different workshops in the 11–12th centuries CE. Moreover, this area could be the origin and the connection with later important glazed ceramic manufacture centers, such as workshops from Teruel and Valencia.

Figure 1. Map of the Iberian Peninsula, showing the approximate border between Islamic and Christian areas at the beginning of the 11th century CE, with some of the main centers of ceramic production.

During the Islamic period, ceramics was one of the most important industries in the Iberian Peninsula due to great trade development, with the production, exportation and importation of ceramic objects. Pottery reflected the influence of Muslim culture, not only in decorative styles but also in decoration technology. As well as a continuity of the production of lead glazes developed in earlier periods, some new technologies were widespread in the *Taifa* kingdoms. The addition of tin oxide to the glazes to obtain opaque decorations was imported from the Middle East and northern Africa in the 10th century CE, or at the end of the 9th century CE. Initially, tin-opacified glazes decorated with green and brown designs on white surfaces were produced in the caliphal Cordoba [4], but the style was fast spread to all the other later kingdoms [4,5]. Another kind of decoration introduced in this period by Muslims was *cuerda seca*. This technique was an important style in al-Andalus, confirmed by the different productions found around the Iberian Peninsula [6,7]; it is characterized by a black line that delimits the drawing of the decoration, and the spaces between lines were filled totally or partly with colored glazes. The geometric and vegetable motifs were common in the techniques of both white tin-opacified glazes and *cuerda seca*.

The continuous Islamic influence in the ceramic production of the Iberian Peninsula was not only present in decoration technology, such as the introduction of tin-opacified glazes and the *cuerda-seca* technique, but also in the decorative styles. All these characteristics and techniques were applied in ceramic production from Albarracin *Taifa* dating from the 11–12th centuries CE, as the numerous objects found during several archaeological excavations have proved. This production was the subject of our study. The research was focused on the technology of production and decoration of the diverse ceramics found in Albarracin from the *Taifa* period.

The aim of this work was the characterization of the Islamic ceramic production of Albarracin in order to establish ceramic reference groups of this period in the *Taifa* and determinate the possible influences of the pottery manufactured in other Islamic kingdoms. To carry out these objectives, the study was focused on the chemical composition of the materials employed in the clay bodies, and on aspects of the decoration related to the composition of the materials and their contribution to the glaze characteristics and to the manufacture process. Finally, this study intended to establish relations between the variations of the raw materials depending on the type of decoration and compare the compositions used in their manufacture with other workshops of the same period.

2. Materials and Methods

Nearly one hundred ceramic fragments, dating to the 11th century CE from archaeological sites in Albarracin [3], all of them decorated, were chosen to carry out the study, although only the results of glazed pottery are discussed in this paper. Most of them belonged to the tableware group (bowls, beakers, and dishes). According to their decoration, the studied samples can be classified in four groups:

- *Cuerda seca* (Figure 2a): ten fragments were studied, decorated only on the main side of the ceramic;
- White tin-opacified glazes (Figure 2(b1,b2)): twenty-two samples with green and brown-black designs on white glazes (Figure 2(b1)) of the main object side were studied. The secondary sides were coated with yellow glazes (Figure 2(b2));
- Yellow/honey glazes (Figure 2(c1,c2)): almost all the ceramics covered with yellow glazes were coated on both sides with the same color. Sometimes these fragments were decorated with green (Figure 2(c1)) or brown lines (Figure 2(c2)) on the yellow glaze. This type of decoration was very frequent, with a great range of colors among the clay bodies. Twenty-two fragments were chosen for the study;
- Green glazes (Figure 2d): fragments decorated with green glazes included ceramics glazed on both sides, with the same green color in many cases, but some fragments had the main side colored in green and the secondary side of yellow color. Nineteen green glazed samples were selected.

Figure 2. Ceramic fragments with different decorations found in Albarracin: (**a**) *Cuerda seca*; (**b1**) White tin-opacified glaze; (**b2**) Yellow glaze on secondary-side of white-tin opacified ceramic; (**c1**) Yellow/honey glaze with green line; (**c2**) Yellow/honey glaze with brown line;(**d**) Green glaze.

Chemical analysis of major and minor elements of the ceramic bodies was carried out in a Thermo Elemental Iris Intrepid Radial Inductively Coupled Plasma-Optical Emission Spectrometer (ICP-OES) (Thermo, Bremen, Germany). Powdered samples were extracted from ceramic bodies by drilling in freshly fractured surfaces of the sherds, using a diamond-tip drill. Samples (50 mg) were digested in open Teflon-vessels. The procedure consisted of successive additions of HNO_3 + HCl, and HF, and a final addition of HNO_3 + $HClO_4$, heating in every step [8]. After dissolution, every sample was diluted in a 50 mL volumetric flask. For ICP-OES analysis, an 1150 W RF power was used. The following components were determined: Na, Mg, Al, K, and Ca as major elements, and Ti, Mn, Fe, Ba, and Sr as minor elements. Results were submitted to a multivariate statistical treatment by hierarchical clustering analysis (using XLSTAT software, version 2021.2.2).

Small fragments of glazed ceramics were cut perpendicularly to the glaze-body interface to prepare polished cross-sections in order to examine them by Optical Microscopy (OM) and Scanning Electron Microscopy (SEM). Cross-sections of samples were observed by means of OM, using a Nikon Eclipse 50ipol (Nikon, Melville, NY, USA). For the determination of glaze composition, a JSM 6360 LV SEM (JEOL, Tokyo, Japan) equipped with an energy-dispersive X-ray analysis (EDS, Inca software, Oxford Instrument, Oxford, UK) with ZAF correction was used, applying an acceleration voltage of 20 kV (oxygen was not measured, oxides were calculated by stoichiometry). The cross-sections were previously carbon coated. X-ray analyses were carried out by scanning large areas of the glaze matrix (normally using ×2000 magnification, although sometimes ×5000 magnification was used where the decoration layer was very thin) and acquiring data from ten different points in every sample.

A Perkin Elmer SCIEX Elan 6000 Inductively Coupled Plasma-Quadrupole Mass Spectrometer (ICP-QMS) (Perkin Elmer, Ontario, Canada), equipped with a cross-flow nebulizer, was used to measure lead isotope ratios in glazes [9]. The measured masses were 204, 206, 207, 208, and 202 to avoid ^{204}Hg interferences.

3. Results and Discussion

3.1. Body Compositions

Chemical compositional data of ceramic bodies obtained by ICP-OES were treated statistically by hierarchical clustering analysis using eight variables (Na, Mg, Al, K, Ca, Ti, Mn, and Fe) and Euclidean distances between samples (Ward method) to group them. This multivariate statistical treatment (Figure 3) revealed several differentiated groups whose compositions are summarized in Table 1 (oxides were calculated by stoichiometry). Calcium was one of the elements whose content varied between groups and, together with other components, allowed the characterization of each group. The groups presented calcareous matrix whose composition ranged from 5 to 18% CaO. Group 3 presented lower average calcium proportion (7.5% CaO) than the other calcareous matrixes, but potassium, magnesium, and aluminum contents were higher. The other two groups with calcareous bodies (group 1 and 2) had similar calcium quantities, but different minor element contents. All studied decorated ceramics were produced with calcareous clay, only a less frequent group of slip-ware decorated with red/brown painted designs (not included in this paper) was made with non-calcareous clay (0.5–1% CaO) [10]. The use of calcareous clays for manufacturing glazed pottery was expected in order to obtain light-colored bodies [11].

The reference groups obtained by cluster analysis, according to the chemical composition of the bodies, could be related to the decoration types manufactured in Albarracin. Groups 1 and 2 corresponded to ceramics decorated with white or green glazes. Most of the *cuerda-seca* fragments were included inside group 1. Group 3 was mainly formed by honey-glazed objects, although some yellow fragments were also included in groups 1 and 2. This fact suggested that yellow-glazed ceramics could be manufactured with different raw materials or in different workshops.

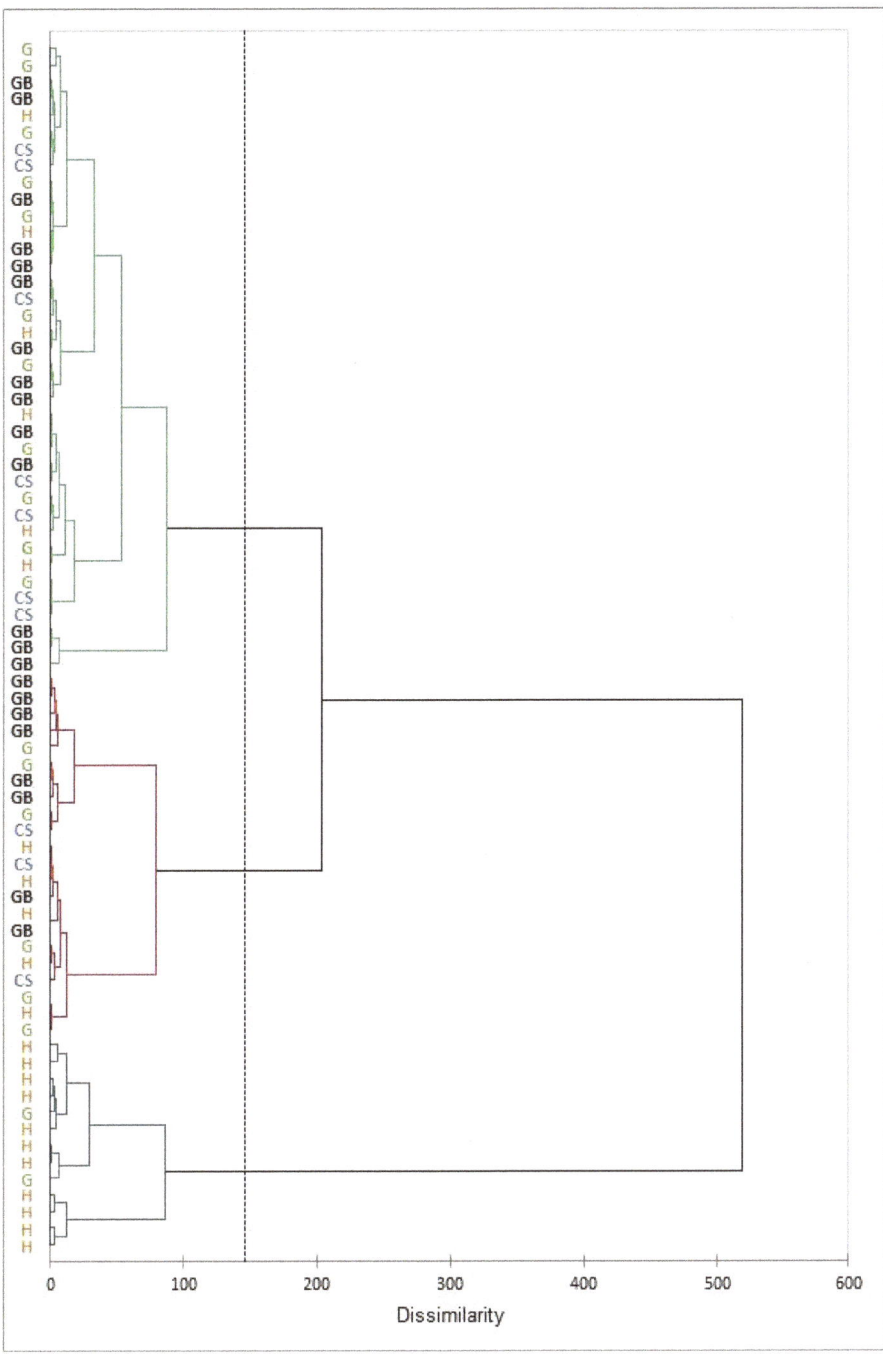

Figure 3. Dendrogram from hierarchical clustering analysis of the body composition data (CS: *cuerda seca*; GB: green and brown decoration on white glaze; G: green glaze; H: yellow/honey glaze).

Table 1. Chemical composition of ceramic bodies, determined by ICP-OES (data as %wt, except Ba and Sr results given as µg.g^{-1}, s: standard deviation; n: number of samples).

		Na$_2$O	MgO	Al$_2$O$_3$	K$_2$O	CaO	TiO$_2$	MnO	Fe$_2$O$_3$	Ba	Sr
Group 1	mean	0.845	2.91	14.8	3.31	12.6	0.587	0.0545	4.78	544	446
	s (n = 38)	0.453	0.94	1.9	0.88	1.4	0.087	0.0136	0.66	152	247
Group 2	mean	0.898	3.56	15.6	3.03	16.2	0.636	0.0643	5.65	597	509
	s (n = 22)	0.495	0.71	1.8	0.67	1.4	0.062	0.0150	0.92	105	148
Group 3	mean	0.734	5.28	16.20	4.70	7.54	0.593	0.0626	5.34	801	393
	s (n = 13)	0.466	0.97	2.39	0.73	2.25	0.107	0.0116	1.31	232	90

According to chemical results, it seems that selection of the raw materials was performed depending on the decoration. In groups 1 and 2, it can be assumed that the same type of illitic clay was employed for the manufacture of the pottery. However, the raw materials used in the ceramic production of group 3 differed in the chemical composition from those employed for groups 1 and 2. Therefore, ceramics included in this last group can belong to another production workshop or even another area.

Comparing the results obtained from Albarracin ceramics with analysis of clay bodies of Islamic productions from other areas of the Iberian Peninsula [4,12–16], it can be pointed out that all the clays used in decorated ceramics were also Ca-rich (8–22% CaO) and had similar contents of iron (4–6% Fe$_2$O$_3$) and fluxing agents (4–5% Na$_2$O+K$_2$O). The use of calcareous clays for decorated pottery justified the buff colors of the bodies due to the growth of calcium silicates, which incorporated some iron atoms to their structure and decreased the content of iron oxides, responsible for red color [11]. A great similarity was found when we compared these objects from Albarracin with other pottery productions from Muslim centers, such as Cordoba, Denia, Granada, Mallorca, Murcia, Pechina, or Zaragoza [4,13–16]. For instance, the Islamic production from Zaragoza during the 11–12th centuries CE was studied [13] and four reference groups were described, based on the type of body clays employed. In these workshops from Zaragoza, differences among the calcareous groups were also based on the amounts of alkalis and the groups were related to the final decoration and to the utility of the pottery. The selection of the raw materials implied great knowledge of their characteristics and properties.

In the south of al-Andalus, the production of ceramics was very popular with a great number of workshops. Cordoba, Denia, Granada, Murcia, and Pechina were important pottery centers during the 11th–13th centuries CE [4,14,16]. The body composition of these glazed productions from Islamic workshops was very similar to ceramics from Albarracin with high contents of calcium (groups 1 and 2), although ceramics manufactured in some of them were richer in calcium content than Albarracin. The study carried out in the workshop of San Nicolas [14], situated in Murcia and dating to the 10th century CE, revealed some differences with the pastes from Albarracin, but three different references groups can be also distinguished; these three groups were also related to the decoration to be applied. Again, this fact demonstrated the specialization of the potters on the selection of the material.

3.2. White Tin-Opacified Glazes

Different types of decoration on the Islamic pottery from Albarracin *Taifa*: white tin-opacified glazes, colored glazes, and *cuerda seca* were studied for their characterization. In the case of white tin-opacified glazed samples, the compositions of white glazes on the main side and the yellow-honey glazes on the secondary side of these fragments were studied. Further, brown and green decorations on the white glazes were analyzed in most ceramic samples. Tin-opacified glazes presented an average chemical composition that was very homogeneous in the main side, as can be seen in Table 2. The major elements found in the white glazes were silicon (SiO$_2$ 42.8%), lead (PbO 36.2%), and tin (SnO$_2$ 7.0%), mixed with contents below 5% of aluminum, iron, alkalis (potassium and sodium) and calcium oxides. Glazes produced during the medieval period in the Iberian Peninsula were characterized by lead-rich contents. The low percentage of aluminum (Al$_2$O$_3$ < 4.5%)

could be attributed to the use of small amounts of clay to produce the glaze, or also to the diffusion of aluminum from the paste to the glaze during firing [17]. Iron and calcium were not added to the recipes of glazes as main components. Their very low proportions (FeO < 1% and CaO 2.86%) were also due to the migration of iron and calcium into the glazes. Higher contents of iron would induce yellow or brown colors in the glaze. Therefore, raw materials employed in the production of tin-opacified glazes were carefully chosen in order to obtain a white color of glaze.

Table 2. Chemical composition (%wt) of white tin-opacified glazes and their colored decoration by EDS.

Glaze Color		Na_2O	MgO	Al_2O_3	SiO_2	K_2O	CaO	MnO	FeO	CuO	SnO_2	PbO
white	mean	1.13	0.60	4.20	42.8	3.48	2.86	-	0.76	-	6.95	36.2
	s (n = 18)	0.27	0.18	0.64	2.2	0.77	0.65		0.29		1.72	2.8
green	mean	1.13	0.56	4.10	42.5	2.89	3.38	-	0.96	2.04	4.53	37.4
	s (n = 10)	0.21	0.17	0.68	2.6	0.56	1.11		0.31	0.66	2.10	4.0
brown	mean	1.10	0.60	3.98	41.1	2.81	3.64	2.83	1.00	1.15	5.06	36.6
	s (n = 18)	0.19	0.21	0.69	2.5	0.70	0.96	1.34	0.41	0.39	1.88	3.9

Green and brown decorations on tin-glazed samples from Albarracin *Taifa* did not differ in their compositions from the white ones, except for the presence of metallic oxides responsible for the colors (Table 2). Analyses carried out on green areas were rich in copper oxide, and brown zones had higher contents of manganese oxide. The content of CuO was in the range of 1.4–3.4% and MnO varied between 1.2 and 6.0%. The presence of copper in brown decorations can be explained due to the thin brown design that made it difficult to determinate only the composition of brown areas without including green zones.

Opacity was accomplished by the presence of small crystals of cassiterite (SnO_2) within the glaze that reflected and scattered the light [18]. These crystals were achieved by the addition of tin oxide to the raw materials of the glazes. During the heating (above 700 °C) of the mixture, the tin oxide was dissolved and began to recrystallize as cassiterite [19]. The amount of tin, the size of tin oxide crystals, their distribution in the glaze, and the thickness of the glaze were factors with great influence on the grade of opacification [12,18,19] and they were related to the process of manufacture. The average amount of tin oxide (7% SnO_2) employed in the studied samples was in good agreement to obtain enough opacity (5–10% SnO_2) without increasing too much the price of the product. The size of tin oxide crystals in Albarracin samples varied from 500 to 800 nm. These dimensions of the crystals indicated that high temperatures (above 800 °C) were reached during the firing process and the cooling occurred fast, as it was demonstrated in the literature [19]. The homogeneous distribution of the cassiterite crystals suggested that the tin glazes were prepared by fritting the raw materials before applying them to the ceramic body [15,20]. Fritting was an extended method used since ancient times to obtain glasses and glazes, which consisted in melting the raw materials (lead, tin, and silica) previously to use them as components of the glazes. Finally, these fired materials were ground to powder before preparing the glaze suspension that was applied on the surface of the body. Besides the homogeneity of the fritted material, frits presented some advantages with respect to raw materials: they reduced the risk of contraction, avoiding cracking the surface, improved the maturation of the glaze, avoided bubbles, and allowed better control of the process [19].

Only one tin-opacified glazed sample had also white glaze on the secondary side; the others had honey glazes on the secondary side, as it has been explained before. The composition of these secondary glazes revealed a higher heterogeneity between specimens (Table 3), but it was possibly to classify them in two different subgroups. The first one included samples with contents of SiO_2 and PbO about 37% and 46% respectively, and the other group was composed by glazes with higher SiO_2 (45%) and lower PbO (37%) proportions. In both cases, the elements added as fluxes were lead and some alkalis with contents around 3–4%. Tin oxide was not present and the proportions of aluminum and iron oxides were increased regarding main tin-opacified glazes, which evidenced the use of clays and sand not as pure as those used to manufacture white glazes. The honey/yellowish color of the glazes was achieved by iron (>2% FeO) dissolved in the glaze.

Table 3. Chemical composition (%wt) of glazes coating the secondary side of white glazed objects by EDS.

Honey Glazes		Na_2O	MgO	Al_2O_3	SiO_2	K_2O	CaO	FeO	PbO
Subgroup 1	mean	0.45	0.81	5.11	36.6	2.57	5.20	2.27	46.5
	s (n = 8)	0.10	0.18	0.52	0.8	0.45	0.74	0.33	2.0
Subgroup 2	mean	0.58	0.93	5.54	44.5	3.62	4.94	2.30	36.9
	s (n = 9)	0.20	0.43	1.74	5.6	0.80	2.09	0.99	3.6

Microstructure of glazes observed by SEM differed depending on the fragment side (Figure 4). Tin-opacified glazes on the main side of the objects (Figure 4a) had few inclusions, while on the secondary side (transparent glazes) were abundant inclusions with large sizes (often bigger than 60 μm) and well distributed through the glaze (Figure 4b,c). In both cases, the inclusion compositions were the same: potassium feldspars and quartz. Those microstructural differences reflected that the materials for the main decoration were carefully chosen and probably fritted for producing the main glazes [14].

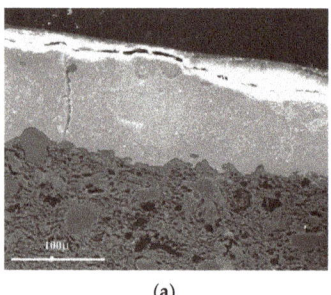
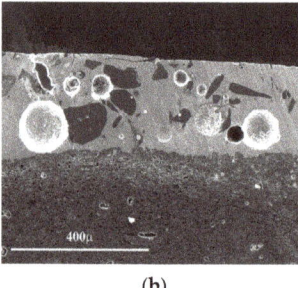

(a) (b) (c)

Figure 4. Tin-opacified glazed ceramics: (a) Secondary electron-SEM image of the main-side glaze with cassiterite crystals scattered in the glaze; (b) Secondary electron-SEM image of the secondary side with non-plastic inclusions; (c) Optical Microscopy image of the secondary-side glaze.

Concerning thickness, it is outstanding that glazes of the main side were thicker than the coatings applied on the secondary side in all studied samples. On average, the thickness of the layers ranged from 100 to 200 μm for tin-opacified glazes, and from 50 to 100 μm for transparent glazes.

In the contact zone between clay body and glaze, several reactions occurred during firing that favored the migration of atoms between both phases. Diffusion of the elements produced the growth of Pb-rich feldspars crystals. The development of these crystals in the interface and the reaction degree between clay body and glaze were related to the type of clay body and the firing/cooling process [17]. Not only in tin-opacified glazes from Albarracin but also in the transparent ones, the amount of crystals and the interaction between clay paste and glaze (interface thickness < 10 μm) was minimum. It was described that, when the glaze was applied on an already fired clay body, the diffusion between both phases was reduced, and consequently the number of crystals formed and the thickness of the interface were lower [17,21]. This feature indicated that tin-opacified glazes manufactured in the *Taifa* of Albarracin were applied on a previously fired paste (biscuited) and a second firing process was carried out to produce the decoration.

If the results obtained for tin-opacified glazes of Albarracin production are compared with other workshops of the Islamic period in the Iberian Peninsula [4,14], it can be pointed out that the technology (firing process and opacity) seemed to be very similar and the glaze recipes differed only slightly. In all Iberian production centers, white opaque glazes were composed by a mixture of lead–silica with tin oxide contents from 5 to 10%. Each workshop had a characteristic lead/silica ratio that could help to differentiate the productions. Islamic ceramic centers in the south and the SE of the Peninsula had higher contents of lead (45–60% PbO and 30–42% SiO_2, in general; then, PbO/SiO_2 ratios from 2 to 1.1) than Albarracin (PbO/SiO_2 = 0.8) and Zaragoza samples (in the NE), except for some examples from the Vega of Granada [16]. The low lead contents of these northeastern groups were compensated by higher amounts of alkaline components (2–5% K_2O); this fact allows the definition of these tin-opacified glazes as potassium-lead glazes. These differences were also revealed in the results obtained with the same methodology in our laboratory in samples from Pechina (9th–10th centuries CE), given by the Laboratoire de Céramologie (Lyon, France) (unpublished data). Those characteristics proved that white tin-opacified glazes manufactured in Albarracin and Zaragoza *Taifas* could be differentiated by their chemical composition from other production areas.

3.3. Monochrome Colored Glazes

Colored glazes were used for coating always both ceramic sides. Yellow/honey glazed samples were decorated on both sides of the objects with the same color; however, green glazed fragments were decorated with either green or yellow glazes on the secondary side. Compositional analyses of green and yellow monochrome glazes (Table 4) revealed that they were rich in lead and silicon, with minor quantities of iron, aluminum, calcium, and alkali oxides. Similar chemical compositions were determined on the secondary sides. Only slight compositional differences were observed among green and yellow glazes. Some of these differences were due to elements responsible of the color of glazes. Although the final color of glazes was also influenced by the body color and the metallic oxide added into the glaze [12,14,17], the latter was the responsible of the compositional differences. Therefore, yellow/honey glazes had higher amounts of FeO (1.6–4% in yellow glazes, versus 0.6–1.7% in green glazes) and green coatings were richer in CuO proportions (1.2–4% in green glazes).

Table 4. Chemical composition (%wt) of monochrome glazes by EDS.

Glaze Color		Na_2O	MgO	Al_2O_3	SiO_2	K_2O	CaO	FeO	CuO	SnO_2	PbO
honey	mean	0.54	0.94	5.00	37.8	2.10	5.03	2.59	-	-	45.1
	s (n = 19)	0.24	0.29	0.72	1.4	0.46	1.44	0.74			3.7
green	mean	0.94	0.74	4.53	40.4	2.84	3.98	1.17	2.56	1.80	41.0
	s (n = 17)	0.36	0.21	0.79	3.5	0.78	1.05	0.35	0.85	1.71	4.5

An important feature of these monochrome glazes was the fact that all green glazes included tin oxide in their composition (Table 4). Tin was detected with EDS analysis and tin-oxide crystals were observed in the green glazes. Although the tin-oxide percentages were lower than white tin-opacified glazes (Table 3), the proportion was enough to give a certain opacity. This characteristic was also highlighted in green glazes from Zaragoza *Taifa* [22].

Inclusions were occasional in the main monochrome glazes (Figure 5), but they were always present in the secondary sides, which implied careful treatment of raw materials again, especially for the main sides. These inclusions were mainly quartz and feldspars. It is not easy to establish an average size of the inclusions because there was a lot of variation even inside the same specimen. The glaze thickness was quite constant in the same sample, but it varied significantly among different ceramic fragments. Additionally, not only the yellow/honey glazes but also the green glazes were thicker in the main side (100–150 µm for the yellow glazes, and 150–200 µm for the green glazes) than in the less important side (80–100 µm in the yellow glazes, and 100–150 µm in the green ones).

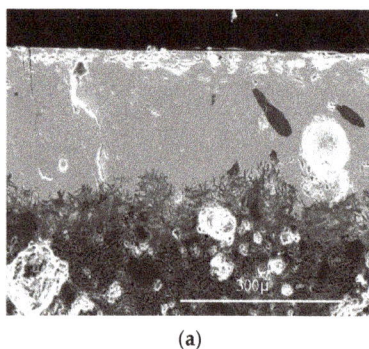 

Figure 5. Monochrome colored glazes: (**a**) Secondary electron-SEM image of monochrome coating where a great interface between glaze and ceramic body can be observed; (**b**) Optical Microscopy image of a green glaze.

In these types of glazes, greater interactions between body and glaze were observed by SEM (Figure 5a). Since the reaction between ceramic paste and glaze was related with the type of developed firing process [12,17,21], it can be assumed that the thickness of the interface (35–40 µm) in monochrome glazes corresponded to a single firing process. This technique of manufacture justified the great amount of lead-potassium feldspar crystals that existed in the interface, and that favored the adhesion between both layers [23]. In addition, this type of firing process enhanced the diffusion of certain elements (such as aluminum and iron) from the body to the glaze. Aluminum and iron were incorporated into the glaze and helped the stabilization of $PbO-SiO_2$ glazes. Therefore, the presence of aluminum and iron in the glazes can be attributed, on the one hand, to the raw materials of the glazes and, on the other hand, to the diffusion of components from the clay to the glaze during the firing process [12,17].

It can be pointed out that the composition of transparent yellow/honey glazes (Table 4) was very similar to that of the glazes employed on the secondary side of white glazes, especially to subgroup 1 (Table 3). Inclusions were abundant in both secondary-side glazes of yellow-honey glazed and white glazed ceramics.

The comparison of chemical compositions of the honey/yellow glazes from Albarracin with productions from other Islamic workshops [4,12,16,22] confirmed differences in lead/silicon ratios. Almost all productions whose workshops were situated in the south and SE part of the Iberian Peninsula had lead oxide content between 45–60% PbO. Lead percentages of glazes produced in Albarracin were among the lowest compared with the rest of the workshops. However, productions from Albarracin and Zaragoza *Taifas* had in common that their ceramics were manufactured with a unique firing process, whereas the rest of the centers of production seemed to prepare monochrome glazes on biscuited ceramics.

3.4. Cuerda-Seca Decoration

Concerning *cuerda-seca* samples, we determined the composition of the black line which divided different areas of the drawing, and the glazes applied inside those areas.

The analyses of the black lines showed inhomogeneous compositions (Table 5) that made it not realistic to establish average proportions. However, the high content of SiO_2 as major element in all the samples is remarkable, mixed with lead, manganese, aluminum, and iron. Manganese was the responsible for the black color of the line, probably in the form of different oxides. Lead could be added to enhance the adherence of the pigments to the paste [24], because the black line was applied directly on the non-fired paste, as can be deduced by the thickness of the interface (Figure 6).

Table 5. Chemical composition (%wt) of the black line, yellow and green glazes in *cuerda-seca* decorated ceramics by EDS.

Glaze Color		Na₂O	MgO	Al₂O₃	SiO₂	K₂O	CaO	MnO	FeO	CuO	SnO₂	PbO
black	interval (n = 8)	0.4–3.4	0.4–9.0	2.3–19	8.5–58	0.5–13	2.2–26	1.3–50	0.6–21	-	-	0.3–30
green	mean	1.00	0.68	3.89	46.0	2.65	3.14	-	0.96	2.29	1.67	37.5
	s (n = 5)	0.22	0.39	0.74	1.7	1.17	0.76		0.38	0.65	1.00	5.1
yellow	mean	0.80	1.22	6.26	38.9	2.04	7.11	-	2.70	-	-	37.9
	s (n = 3)	0.14	0.46	1.25	5.4	0.67	2.02		0.03			11.9
turquoise	mean	2.47	1.09	1.36	50.6	2.90	4.21	-	0.56	1.86	11.6	23.3
	s [1] (n = 1)	0.19	0.15	0.47	2.2	0.23	0.43		0.21	0.16	3.4	1.4

[1] In this case, standard deviation (s) corresponds to the number of analyzed points (9) in one sample.

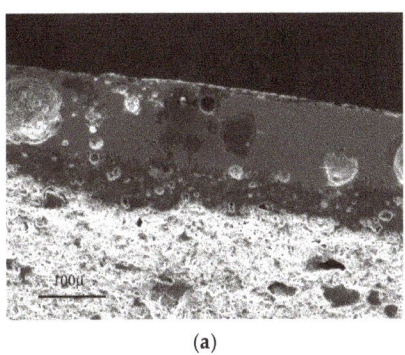 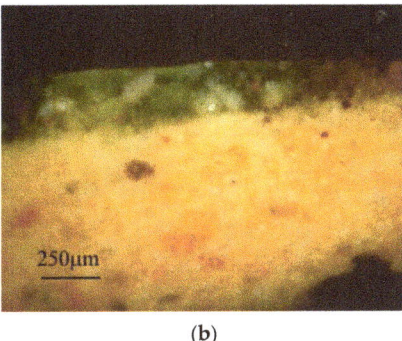

(a)   (b)

Figure 6. *Cuerda-seca* decorated glazes: (**a**) Secondary electron-SEM image of the glaze; (**b**) Optical Microscopy image of *cuerda-seca* decoration.

The glazes used to fill the spaces between black lines were rich in silicon and lead (Table 5). The green glazes had tin oxide in lower proportions than white tin-opacified glazes. Although there were not many *cuerda-seca* samples with yellow and green glazes among the studied samples, and the yellow ones showed great variation in lead content, the average composition of the PbO was similar to the secondary yellow/honey glazes (subgroup 2, Table 3) on green-and-brown decorated objects, and to green monochrome glazes (Table 4). However, the silicon proportion of *cuerda-seca* glazes differed more in both colored glazes.

Once again, the presence of copper and iron oxides with percentages around 2–3% contributed to the final color of the glazes. Only a sample showed *cuerda-seca* decoration with honey and green-turquoise colors. In this case, the color was due to the presence of copper, but its green hue was shifted to turquoise because the glaze had alkaline-lead composition with a significant contribution of sodium (2.5% Na₂O) and potassium (2.9% K₂O) and lower lead content (23.3% PbO). These differences in green hues were already highlighted in *cuerda-seca* ceramics of the 12th century CE produced in the Iberian Peninsula [25].

Studies of pottery with *cuerda-seca* glazes manufactured in the Iberian Peninsula during the 10th–12th centuries CE revealed the use of silicon and lead as main components [6,24,25]. The amounts of lead decreased from the 10th century (43–55% PbO) to 12th century CE (34–43% PbO), mainly in productions from the south of al-Andalus [25]. Nevertheless, samples manufactured during the 11th century CE in the NE workshop of Zaragoza were characterized by high lead proportions [24]. Contrary to other types of decoration, pottery decorated with *cuerda seca* produced in Albarracin in the 11th-12th centuries CE was more similar to the recipes employed in the south of the Peninsula with lead contents around 35–38% PbO.

Examinations by SEM of polished sections of *cuerda-seca* glazes confirmed the presence of abundant inclusions scattered homogeneously in the glaze (Figure 6). The inclusion sizes varied from 50 to 70 μm and their composition was mainly quartz. The use of such an amount of inclusions and their size, plus the tin-oxide content, could be responsible for the opacity. Opacification can be reached either with undissolved inclusions into the glazes or with tin-oxide crystals scattered in the glaze, and even with both of them [26,27]. *Cuerda-seca* ceramics manufactured in the same period in al-Andalus employed both techniques [6,24,25], although only tin oxide was employed as an opacifier in later centuries. Therefore, the inclusions found in the glazes played an important role in the opacity of the glazes. The thickness of the glazes reached values higher than 200 μm and the thickness of the clay-glaze interfaces was also higher (>45 μm), which implied that a single firing technique was carried out during the pottery's manufacture. The type of firing process (single or double) depended on the workshops; both techniques were used without a clear tendency. Single firing process was more frequent, especially in workshops such as Almeria and Zaragoza [24,25]. This characteristic was also followed in the studied production from Albarracin *Taifa*.

As it has been explained before, the technology of opacification also underwent an evolution from a double method (tin oxide and inclusions) employed in the 10th century CE to the opacity technique due exclusively to tin oxide used in the 12th century CE. However, in our case, the double methodology was followed, as in the case of the southern workshops.

It seems clear that the *cuerda-seca* technique underwent an evolution during the Islamic period with variations in recipes, firing processes, and methods of opacification. The evolution did not follow a regular extension to all the production centers, but we can highlight that Albarracin was influenced by the southern centers of al-Andalus. Although archaeological studies based on the decoration motifs of *cuerda seca* during the period of *Taifas* emphasized that the same type of vegetal designs was also employed in Zaragoza [6], both influences could be possible taking in mind the good relationships with northern and southern *Taifas*, and the wide commercial distribution of Islamic glazed ceramics [28,29].

3.5. Lead and Raw Materials for Glazes

Lead isotopes ratios were measured by ICP-QMS in a selection of representative glazed samples, because of the important information in lead sources that could be provided. Natural lead consists of four isotopes: three of them (^{208}Pb, ^{207}Pb and ^{206}Pb) come from the radioactive decay of U and Th, but the fourth one (^{204}Pb) has a non-radiogenic origin and natural lead changes its isotopic composition. The results of twenty-three samples from Albarracin, plus two samples from Islamic workshops of Zaragoza and one sample from Pechina are shown in Figure 7.

The lead-isotope ratios of the samples, plotted as ^{208}Pb/^{206}Pb vs. ^{207}Pb/^{206}Pb and ^{206}Pb/^{204}Pb vs. ^{207}Pb/^{206}Pb ratios (Figure 7), offered the possibility of establishing several groups and distinguishing different features of the used lead for the production of glazes. Samples from Albarracin can be clearly separated into two groups. One of the groups was related to the fragments with yellow/honey glazes (Figure 7a,b, on the right of both plots). The second group was formed by white tin-opacified and green glazes from Albarracin (Figure 7a,b, on the left of both plots). Moreover, these last two types of glazes showed slight differences between them, but both had in common being tin-opacified glazes. If we observed the results of the three samples from other areas (Zaragoza and Pechina), the white tin-opacified glaze from Pechina was grouped with the tin-opacified samples from Albarracin. However, the glazes from Zaragoza indicated differences in their source of lead, because both samples were closer to the honey glazes from Albarracin, and even the white tin-opacified glaze from Zaragoza (marked with a red circle in Figure 7a,b) was unexpectedly similar to the transparent honey glazes from Albarracin.

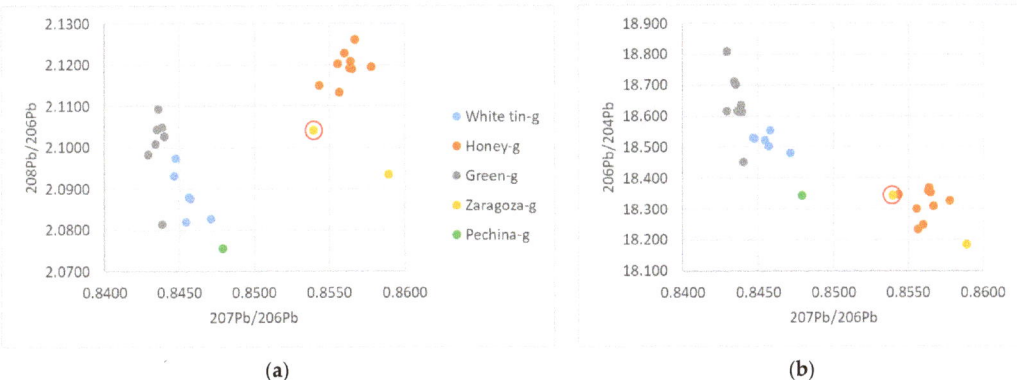

Figure 7. Lead-isotope ratios of glazes from different ceramics (white tin-opacified glazes, honey glazes, and green glazes from Albarracin; glazes from Zaragoza workshops (white tin-opacified glaze is marked with a red circle); and glaze in a sample from Pechina): (**a**) $^{208}Pb/^{206}Pb$ ratio vs. $^{207}Pb/^{206}Pb$ ratio; (**b**) $^{206}Pb/^{204}Pb$ ratio vs. $^{207}Pb/^{206}Pb$ ratio.

Therefore, all of these glazed ceramics, manufactured in Islamic Albarracin, had lead isotope abundances very well characterized in two groups related to the type of glaze (tin-opacified or transparent) applied on the ceramics. This suggested diverse lead sources for the glaze raw materials depending on the type and quality of glaze; then, it could be related to the use of a type of commercial lead for producing tin-opacified glazes and another different lead product for preparing transparent glazes. The first one could be linked to obtaining a best-quality glaze or to the preparation (fritting) of tin-lead glazes. In later periods (16th century CE), differences in the commercial lead supplied to potters were documented: it was explained that the lead used to produce glazed cooking pots was cheaper than the lead acquired to prepare tin-opacified glazes [30].

It is important to underline that this possible source or supplier of lead for tin-opacified glazes from Albarracin was quite similar to that for the white glaze from Pechina; as such it could perhaps come from one of the lead sources in the south/southeast of the Iberia Peninsula. However, lead used in Zaragoza workshops seemed to be different, or closer to that introduced in transparent glazes; slight differences could be highlighted between these two samples (honey and white) (Figure 7a,b), but there are not enough results to find significant dissimilarities.

4. Conclusions

The results of this study allowed the characterization of the pottery production from Albarracin *Taifa* in the 11–12th centuries CE, and establishing differences and similarities with Islamic productions manufactured in other *Taifa* kingdoms of the Iberian Peninsula during the same period. The analyses carried out on the ceramic bodies confirmed that the selection of raw materials depended on the decoration to be applied afterwards. All the samples had a Ca-rich body in order to obtain buff colors and to avoid cracking during firing process.

Firing process depended on the kind of decoration. Tin-opacified glazes were applied on pre-fired clay bodies, but *cuerda-seca* and transparent yellow/honey glazes were applied on the unfired clay body. Once again, this differentiation between the two techniques implies a specialization of the potters.

Concerning to the decoration, it can be pointed out some conclusions about Albarracin production. Glazes manufactured in Albarracin *Taifa* during 11th century CE were rich in lead and silicon oxides. Their composition ranged from 32% to 48% PbO and from 36 to 49% SiO_2, but differences could be established between white tin-opacified and colored glazes. Secondary-side glazes had similar compositions, although the number of inclusions and the size of them were in both cases bigger. However, the thickness of the coatings was thinner on the less important side. Those features corroborated the idea of a specialization and knowledge of the techniques of production.

As well as silicon and lead as main components, tin-opacified glazes included cassiterite crystals that were responsible for opacity. The buffer color of the bodies required only ~7% SnO_2. Tin-oxide crystals were well distributed inside the glazes, and the size of them (<800 nm) was an important factor to achieve the opacification. Green monochrome glazes were always opacified with tin oxide. However, opacity in *cuerda-seca* samples was reached by the combination of tin oxide plus undissolved grains of quartz. This technique of opacification was also used in the south of Iberian Peninsula in the 10th century CE.

Copper, iron, and manganese oxides were added to the glazes to obtain, respectively, green, yellow, and black glazes. Manganese oxide was employed to draw the black lines that described the designs of *cuerda-seca* specimens.

Comparing the lead/silicon ratios used in the manufacture of ceramics with transparent and tin-opacified glazes in Albarracin *Taifa* with other production centers of the Iberian Peninsula during the Islamic period, we can conclude that PbO/SiO_2 ratios were lower in the northeastern workshops than in the southern ones. However, the potassium content was higher in the northeastern centers.

In all the studied ceramic workshops of the Iberian Peninsula, tin-opacified glazes were applied on a biscuited body, as in Albarracin *Taifa*. However, there existed differences in transparent glazes, because the southern workshops seemed to use a double firing for producing this type of glaze. In the case of *cuerda-seca* decoration, the production of Albarracin had more similarities in the method of opacity and in the lead/silicon ratios with southern workshops; however, for the firing method, Albarracin samples followed a single firing process, as with the Zaragoza and Almeria centers.

According to these results, it can be highlighted that Islamic ceramics with tin-opacified and transparent glazed decorations manufactured in the *Taifa* of Albarracin during the 11th century CE were influenced by the production of the north *Taifa* of Zaragoza. Nevertheless, ceramic decorated with the *cuerda-seca* technique was influenced in some aspects by southern workshops, and in other ones by centers of production from Zaragoza or Almeria.

These possible influences and the Islamic potters' knowledge and specialization were also revealed by the lead-isotope ratio results, where two different lead sources or suppliers were shown: one to prepare tin-opacified glazes, and another one for yellow/honey glazes.

Author Contributions: Conceptualization, resources, writing—review and editing, supervision, project administration, funding acquisition, J.P.-A.; methodology, validation, investigation, data curation, writing—original draft preparation, P.M. and J.P.-A. All authors have read and agreed to the published version of the manuscript.

Funding: This research was funded by the Comunidad de Trabajo de los Pirineos (Aragon-Catalonia-Aquitaine) and the Diputación General de Aragón (DGA).

Acknowledgments: The authors would like to thank the Fundación Santa Maria de Albarracin and the Museum of Teruel for providing the samples, and Julián Ortega for selecting and describing the archaeological fragments. We would like to acknowledge the use of Servicio General de Apoyo a la Investigación-SAI, Universidad de Zaragoza.

Conflicts of Interest: The authors declare no conflict of interest.

References

1. Almagro, A.; Jiménez, A.; Ponce de León, P. *Albarracin. El Proceso de Restauración de su Patrimonio Histórico*; Fundación Santa María de Albarracin: Albarracin, Spain, 2004; pp. 13–24.
2. Bosh Vilá, J. *Albarracín musulmán. Historia de Albarracín y su Sierra*; Instituto de Estudios Turolenses: Teruel, Spain, 1959; Volume 2.
3. Ortega Ortega, J.M. *Anatomía del Esplendor*; Fundación Santa María de Albarracin: Albarracin, Spain, 2007.
4. Salinas, E.; Pradell, T. The transition from lead transparent to tin-opacified glaze productions in the western Islamic lands: Al-Andalus, c. 875–929 CE. *J. Archaeol. Sci.* **2018**, *94*, 1–11. [CrossRef]

5. Rosselló, G. La céramique verte et brune en al-Andalus du Xe au XIIIe siècle. In *Le vert & le Brun, de Kairouan à Avignon, Céramiques du Xe au XVe siècle*; Réunion des Musées Nationaux: Marseille, France, 1995; pp. 105–117.
6. Déléry, C. Dynamiques Économiques, Sociales et Culturelleres D'al-andalus à Partir d'une Étude de la Cerámique de Cuerda seca (Seconde Moitié du Xe siècle-Première Moitié du XIIIe Siècle). Ph.D. Thesis, Université Toulouse II, Toulouse, France, 2006.
7. Déléry, C. Using cuerda seca ceramics as a historical source to evaluate trade and cultural relations between Christian ruled lands and al-Andalus, from the tenth to thirteenth centuries. *Al-Masaq J. Mediev. Mediterr.* **2009**, *21*, 31–58. [CrossRef]
8. Pérez-Arantegui, J.; Uruñuela, M.I.; Castillo, J.R. Roman glazed ceramics in the Western Mediterranean: Chemical characterization by Inductively Coupled Plasma Atomic Emission Spectrometry of ceramic bodies. *J. Archaeol. Sci.* **1996**, *23*, 903–914. [CrossRef]
9. Marzo, P.; Laborda, F.; Pérez-Arantegui, J. A simple method for the determination of lead isotope ratios in ancient glazed ceramics using Inductively Coupled Plasma-Quadrupole Mass Spectrometry. *Atom. Spectrosc.* **2007**, *28*, 195–201.
10. Marzo, P. Búsqueda y Aplicación de Nuevos Parámetros Analíticos Para el Estudio y la Recuperación del Patrimonio: El Material Cerámico de la Taifa de Albarracín. Ph.D. Thesis, Universidad de Zaragoza, Zaragoza, Spain, 2010.
11. Molera, J.; Pradell, T.; Vendrell-Saz, M. The colours of Ca-rich ceramic pastes: Origin and characterization. *Appl. Clay Sci.* **1998**, *13*, 187–202. [CrossRef]
12. Molera, J. Evoluciò Mineralógica i Interacciò de les Pastes Càlciques amb els Vidriats de plom: Interaccions Arqueomètriques. Ph.D. Thesis, University of Barcelona, Barcelona, Spain, 1996.
13. Lapuente, P.; Pérez-Arantegui, J. Characterization and technology from studies of clay bodies of local Islamic production in Zaragoza (Spain). *J. Eur. Ceram. Soc.* **1999**, *19*, 1835–1846. [CrossRef]
14. Molera, J.; Vendrell-Saz, M.; Pérez-Arantegui, J. Chemical and textural characterization of tin glazes in Islamic ceramics from eastern Spain. *J. Archaeol. Sci.* **2001**, *28*, 331–340. [CrossRef]
15. Salinas, E.; Pradell, T.; Molera, J. Glaze production at an early Islamic workshop in al-Andalus. *Archaeol. Anthropol. Sci.* **2019**, *11*, 2201–2213. [CrossRef]
16. Molera, J.; Carvajal, J.C.; Molina, G.; Pradell, T. Glazes, colourants and decorations in early Islamic glazed ceramics from the Vega of Granada (9th to 12th centuries CE). *J. Archaeol. Sci. Rep.* **2018**, *21*, 1141–1151. [CrossRef]
17. Molera, J.; Pradell, T.; Salvadó, N.; Vendrell-Saz, M. Interactions between clay bodies and lead glazes. *J. Am. Ceram. Soc.* **2001**, *84*, 1120–1128. [CrossRef]
18. Vendrell, M.; Molera, J.; Tite, M.S. Optical properties of tin-opacified glazes. *Archaeometry* **2000**, *42*, 325–340. [CrossRef]
19. Molera, J.; Pradell, T.; Salvadó, N.; Vendrell, M. Evidence of tin oxide recrystallization in opacified lead glazes. *J. Am. Ceram. Soc.* **1999**, *82*, 2871–2875. [CrossRef]
20. Molera, J.; Pradell, T.; Salvadó, N.; Vendrell, M. Lead frits in Islamic and Hispano-Moresque glazed productions. In *From Mine to Microscope: Studies in Honour of Mike S. Tite*; Shortland, A., Freestone, I., Rehren, T., Eds.; Oxbow Books: Oxford, UK, 2009; pp. 1–10.
21. Tite, M.S.; Freestone, I.; Mason, R.; Molera, J.; Vendrell, M.; Wood, N. Lead glazes in Antiquity. Methods of production and reasons for use. *Archaeometry* **1998**, *40*, 241–260. [CrossRef]
22. Pérez-Arantegui, J.; Castillo, J.R. Chemical characterisation of clear lead glazes on Islamic ceramics, produced in northern al-Andalus (Muslim Spain). In *Archaeometry 98, Proceedings of the 31st Symposium, Budapest, Hungary, 1998*; Jerem, E., Biró, K.T., Eds.; BAR International Series 1043, Archeopress, Central European Series 1; Archaeopress: Oxford, UK, 2002; Volume II, pp. 635–639.
23. Molera, J.; Pradell, T.; Martinez-Manent, S.; Vendrell-Sanz, M. The growth of sanidine crystals in the lead glazes of Hispano-Moresque pottery. *Appl. Clay Sci.* **1993**, *7*, 483–491. [CrossRef]
24. Pérez–Arantegui, J.; Soto, M.; Castillo, J.R. Examination of the "cuerda seca" decoration technique on Islamic Ceramic from al Andalus (Spain). *J. Archaeol. Sci.* **1999**, *26*, 935–941. [CrossRef]
25. Chapoulie, R.; Déléry, C.; Daniel, F.; Vendrell-Saz, M. Cuerda seca ceramics from al-Andalus, Islamic Spain and Portugal (10th-12th centuries AD): Investigation with SEM-EDX and Cathodoluminiscence. *Archaeometry* **2005**, *47*, 519–534. [CrossRef]
26. Mason, R.B.; Tite, M.S. The beginning of tin-opacification of pottery glazes. *Archaeometry* **1997**, *39*, 41–58. [CrossRef]
27. Molera, J.; Vendrell-Saz, M.; García-Valles, M.; Pradell, T. Technology and colour development of Hispano-Moresque lead-glazed pottery. *Archaeometry* **1997**, *39*, 23–39. [CrossRef]
28. Beltrame, M.; Santos, J.R.; Gómez Martínez, S.; Correia, F.B.; Candeias, A.; Mirão, J. Nova variante de cerâmica decorada a "verde e manganês" em Évora. *Conimbriga* **2015**, *54*, 225–247. [CrossRef]
29. Bugalhão, J. The Production and Consumption of Islamic Ceramics in Lisbon. *Al-Masaq J. Mediev. Mediterr.* **2009**, *21*, 83–104. [CrossRef]
30. Alvaro Zamora, I. *Cerámica Aragonesa*, 1st ed.; Ibercaja: Zaragoza, Spain, 2002.

Article

Gaming in Pre-Roman Italy: Characterization of Early Ligurian and Etruscan Small Pieces, Including Dice

Ivana Angelini [1,2,*], Cinzia Bettineschi [1], Marica Venturino [3] and Gilberto Artioli [2,4]

1. Dipartimento dei Beni Culturali, Università di Padova, 35139 Padova, Italy; cinzia.bettineschi@unipd.it
2. Consorzio Interuniversitario Nazionale per la Scienza e Tecnologia dei Materiali, INSTM, 50121 Firenze, Italy
3. Archeologa, già Funzionario della Soprintendenza Archeologia, Belle Arti e Paesaggio per le Province di Alessandria, Asti e Cuneo, 15122 Alessandria, Italy; marica.venturino@gmail.com
4. Dipartimento di Geoscienze, Università di Padova, 35131 Padova, Italy; gilberto.artioli@unipd.it
* Correspondence: ivana.angelini@unipd.it

Featured Application: Several unusual small objects from the Villa del Foro archaeological excavation were characterized and interpreted either as gaming pieces or functional materials.

Abstract: An interesting assemblage of ancient ceramic materials connected or potentially connected with gaming activities has been characterized from the archaeometric point of view. The materials (washer-like pieces, small spheres, and cubic dice, with and without inscriptions) were found in the Villa del Foro excavation (Alessandria, Italy). They are related to the early Ligurian population of the site and their frequent contacts with Etruscan both in Etruria and in the Po Valley, in a period spanning the early VI century BC till the first half of the V century BC. Starting from the materials evidence, hypotheses are proposed concerning their possible use and cultural meaning. The studied cubic dice are discussed in the wider context of the pre-Roman diffusion of these objects.

Keywords: dice; gaming pieces; Ligurian; Etruscan; clay spheres

1. Introduction

There is substantial knowledge on ancient games, especially on the use of board games in the ancient cultures of the Middle East (see for example: [1]), to the extent that for some of them, detailed rules for playing and general utilization are sufficiently well understood (Finkel 2007 [1], pag. 16–32), as is their role in social activities and cultural transmission [2,3]. However, investigations on ancient gaming materials in Europe are rather limited. Only in the last thirty years have several thematic exhibitions focused on the games in ancient times and the materials related to the games themselves [4–8].

The recent find of an unusual assemblage of small objects during the archaeological excavation at Villa del Foro (Alessandria, Italy) [9], possibly related to gaming practices in Pre-Roman cultures, has prompted an in depth characterization of the materials as an aid to the interpretation of their manufacturing and use [10,11]. The assemblage is composed of cubic dice, washer-like pieces, and small spheres. The archaeometric characterization of the objects is presented here, as well their interpretation based on the available knowledge of coeval similar objects.

2. Materials and Methods

The analyzed objects are listed in Table 1. They were found during the archaeological excavation in different locations of the site: about one-third are sporadic finds or were recovered on the surface, whereas two-thirds belong to specific stratigraphic units. The archaeological information related to the exact location and stratigraphy of the objects is discussed in detail in Paltineri [10]. Summarizing the results: one washer-like piece and one sphere were recovered from the arable layer; two washer-like pieces, one sphere and

five uninscribed dice are from an alluvial deposit (area B, US 2010). Four uninscribed dice were found in a combustion pit together with abundant ceramic sherds, spindles and animal remains (area E, US 1531); one uninscribed dice and one sphere associated to dice with irregular numbers or signs (Figure 1d) were found in the two filling layers (US 1649 and US 1549) of the same combustion pit. In the same area (area E) other pieces were recovered: an uninscribed dice from a filling layer (US 2122) of a second pit containing ceramic fragments; and an uninscribed dice and a washer-like piece from a dumping (US 1505) possibly related to a production zone. The dumping material also yields spindles, large terracotta rings, numerous ceramic sherds, fragments of grindstones, and furnace remains. A different area of the site (area F) yielded three washer-like pieces from an ancient layer of the settlement (US 1501); a fired surface probably used for food preparation (US 1514) and a fired clay area with unknown function (US 1571) [10]. Even if there are no systematic association patterns between the different types of objects, at least in one case uninscribed dice are in a secure connection with sphere and washer-like pieces.

Based on the ceramic types associated with the finds the proposed date for the uninscribed dice is VI century BC; a washer-like piece and another uninscribed dice are dated to the end of the VI-beginning of the V century BC, and another washer-like piece is dated to the first half of the V century BC. However, the occurrence and distribution of the objects indicate that they were in use during the whole life span of the site, that is between the early VI century BC till the first half of the V century BC [10]. The small objects of interest are mostly composed of fired ceramics, except one dice made out of bone, and they can be described by three principal shapes: (1) cylindrical washer-like discs, (2) small spheres, and (3) cubic dice (Figure 1).

All samples were initially weighed and dimensionally measured with a caliper. The results are reported in Table 1. In the case of irregular shapes, the minimum and maximum dimensions are listed.

Two coeval dice reported from the nearby site of Castello di Annone [12] are also listed in Table 1, they were included because they bear similarities with the Villa del Foro materials and because one of the two dice has specific Etruscan letters. No analytical data are available for these samples, except for the stereomicroscopy investigation of the surfaces. They are made of ceramics as well. One of the two dice is unmarked, whereas the other bears Etruscan letters on two sides, and a variable number of small incised dots (9–12–17–22) randomly distributed on the faces (Figure 2), plus several linear marks on the edges.

Based on macroscopic observations, the major differences between objects are (1) the overall size, (2) the presence or absence of finishing pigmentation on the surface, and (3) the presence or absence of incised decorations and/or inscriptions.

All samples were further characterized by X-ray powder diffraction (XRPD) and Raman spectroscopy, in order to control the composition of the ceramics, the degree of firing, and the composition and application technique of the surface pigments.

The XRPD measurements were made on minute quantities of material using a PANalytical X'Pert Pro goniometer operated in θ-θ Bragg–Brentano geometry and equipped with a Pixcel RTMS detector. Measuring conditions were: Cobalt Kα radiation, 40 kV and 40 mA power, angular range 3–85° 2θ, virtual steps of 0.02° 2θ. Because of the very small amount of material extracted from the objects, the signal was optimized by using a rotating zero-background sample holder. The diffraction spectra were analyzed using the X'Pert HighScore Plus software 3.0 of PANalytical.

Micro-Raman analyses were performed with a DXR Thermo Scientific instrument, equipped with a 532 nm laser and a 50× LWD (Long Working Distance) objective. The working condition selected for the analyses is acquisition time 3 s, 32 scans, 5 mW and a 25 µm pinhole. The spectra were processed with the Omnic 9 software (Thermo Scientific, Waltham, MA, USA) and compared with reference spectra recovered from the online RRUFF database (http://rruff.info) and from our internal database of minerals and pigments.

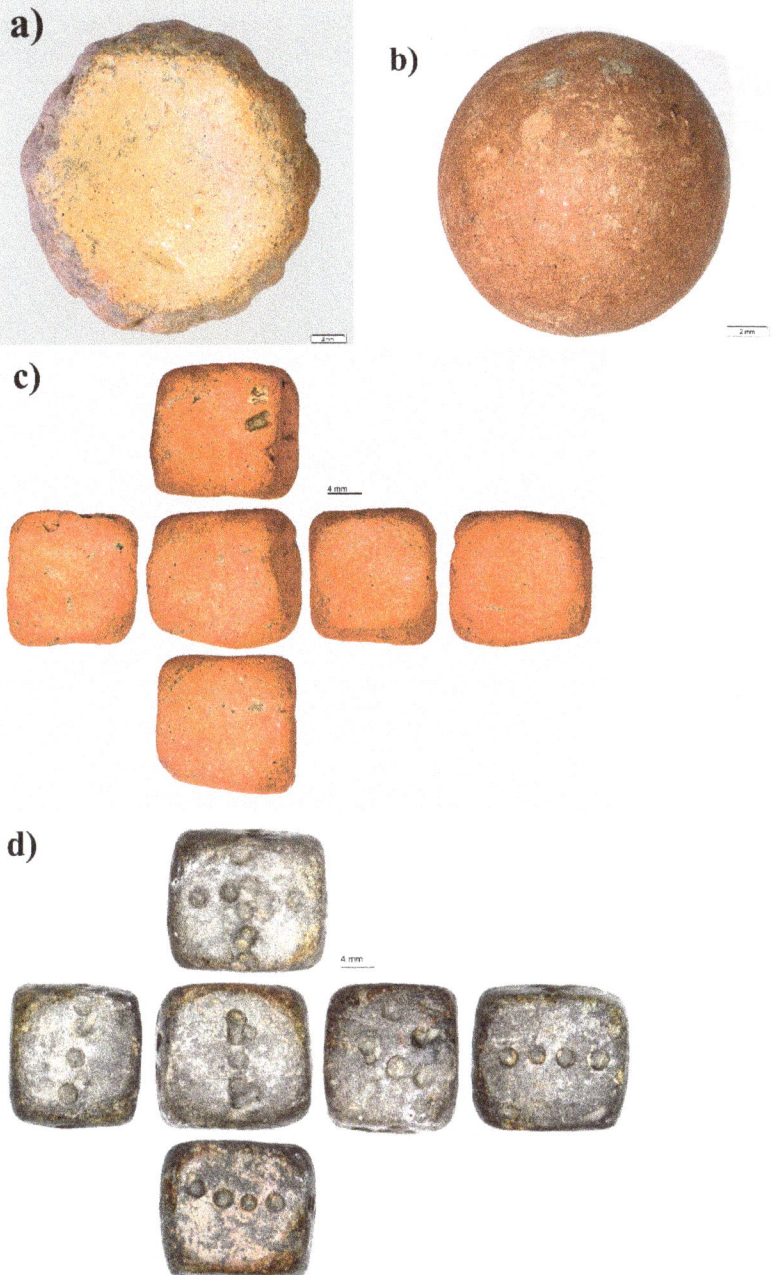

Figure 1. Representative shapes of the investigated objects from Villa del Foro: (**a**) washer-like disc, with indented borders (VF91.157/E1); (**b**) small ceramic sphere with a red colored surface (VFRS111); (**c**) cubic dice with no inscriptions and red-pigmented surface (VF14.SP384D); (**d**) cubic dice with numbers or other signs (VF90.1549/E32). The scale bars are (**a**,**b**) 2 mm, and (**c**,**d**) 4 mm.

Table 1. List of investigated materials from Villa del Foro (Alessandria, Italy).

Id.	Catalog n.	Location	Material	Description	Dimensions (cm)	Weight (g)
				Washer-like discs		
1	VF86.1000.12	area A, S5E10 g7, US 1000	fired ceramics	Irregular clay disc, with rough reeded edge	ø max 2.7; ø min 2.4; h. 0.8 (ø mean 2.5)	5.50
2	VF86.1210.1076	area B, US 1210	fired ceramics	Roughly modeled clay disc, irregular circle shape, flat section	ø max 2.9; ø min 2.8; h. 0.6 (ø mean 2.85)	6.52
3	VF86.1210.917	area B, S6E10, US 1210	fired ceramics	Rough clay incomplete semi-circle, flat section, nail marks on one surface	ø max 5.8; ø min 3.7; h. 1.1 (ø mean 4.85)	23.46
4	VF88.1501.E39	area E, US 1501	fired ceramics	Irregular clay disc, flat asymmetric section	ø max 2.9; ø min 2.5; h. 0.9 (ø mean 2.75)	6.99
5	VF91.1571.E41	area E, US 1571	fired ceramics	Flat circular clay disc, with one slightly concave surface, vertical nail marks on the edge	ø max 2; ø min 1.9; h. 1 (ø mean 1.95)	19.50
6	VF.SP244	stray find	fired ceramics	Irregular flat clay disc	ø max 2.9; ø min 2.8; h. 0.7 (ø mean 2.85)	5.68
7	VF.SP251	stray find	fired ceramics	Irregular clay oval disc, carved out of a pot fragment	ø max 6.4; ø min 5.5; h. 1 (ø mean 5.98)	49.90
8	VF.SP187	stray find	fired ceramics	Roughly modelled fragmented clay emi-sphere	ø max 1.8; ø min 1.6; h. 1.4 (ø mean 1.70)	3.82
9	VF90.1514.E40	area E, US 1514, [US 1530]	fired ceramics	Irregular clay circular disc with eleven engraved lines on a side	ø max 5.6; ø min 5.4; h. 0.7	5.50
				Spheres		
10	VF89.1210-VIII.B24	area B, US 1210 VIII tg.	fired ceramics	Irregular clay sphere	ø max 1.9; ø min 1.7 (ø mean 1.80)	6.02
11	VF90.1549.E33	area E, US 1549 [US 1531]	fired ceramics	Irregular clay sphere	ø max 1.4; ø min 1.3 (ø mean 1.35)	2.98
12	VF.RS111	area C, surface find	fired ceramics	Irregular clay sphere, surface pigmented in red	ø max 1.4; ø min 1.4 (ø mean 1.40)	2.81
13	VF.SP247	part. 199	fired ceramics	Irregular clay sphere	ø max 1.9; ø min 1.7 (ø mean 1.80)	4.86
14	VF14.SP385D	stray find	fired ceramics	Irregular clay sphere	ø max 1.4; ø min 1.2 (ø mean 1.30)	1.94

Table 1. Cont.

Id.	Catalog n.	Location	Material	Description	Dimensions (cm)	Weight (g)
				Dice with no marks or inscriptions		
15	VF88.1210-X.B22	area B, US 1210 X tg.	fired ceramics	Irregular clay cube, rounded edges	1.19; 1.15; 1.14 (side mean 1.16)	1.84
16	VF88.1210-II.B23	area B, US 1210 II tg.	fired ceramics	Irregular clay cube, rounded edges	1.36; 1.33; 1.32 (side mean 1.33)	2.06
17	VF87.1210-III.267	area B, US 1210 III tg.	fired ceramics	Irregular clay cube, rounded edges	1.37; 1.33; 1.24 (side mean 1.31)	3.69
18	VF87.1210-V.49	area B, US 1210 V tg.	fired ceramics	Irregular clay cube, rounded edges	1.63; 1.56; 1.54 (side mean 1.57)	6.34
19	VF87.1210-V.B3	area B, US 1210 V tg.	fired ceramics	Fragmented irregular clay cube, rounded edges	1.28; 1.21; 1.05 (side mean 1.18)	2.01
20	VF07.2122.1D	area M, US 2122 [2080]	fired ceramics	Irregular clay cube	0.99; 0.98; 0.96 (side mean 0.97)	1.53
21	VF90.1000.E36	area E, US 1000	fired ceramics	Irregular clay cube, rounded edges	1.30; 1.31; 1.42 (side mean 1.34)	3.05
22	VF88.1505.E37	area E, US 1505	fired ceramics	Irregular clay cube, rounded edges, blackened surface	1.61; 1.59; 1.58 (side mean 1.59)	6.94
23	VF90.1549.E34	area E, US 1549 [US 1531]	fired ceramics	Irregular clay cube, rounded edges	1.46; 1.44; 1.43 (side mean 1.44)	4.53
24	VF91.1531.E38	area E [US 1531]	fired ceramics	Irregular clay cube, rounded edges, surface pigmented in white	1.84; 1.82; 1.80 (side mean 1.82)	8.60
25	VF90.1649.E35	area E, US 1649 [US 1531]	fired ceramics	Irregular clay parallelepiped, concave surfaces	1.26; 1.23; 0.87 (side mean 1.12)	1.56
26	VF.RS236	surface find N11W1	fired ceramics	Irregular clay cube, rounded edges, surfaces pigmented in red	1.65; 1.60; 1.41 (side mean 1.55)	5.17
27	VF.RS237	surface find N11W9	fired ceramics	Irregular clay cube, rounded edges	1.60; 1.55; 1.43 (side mean 1.52)	5.79
28	VF.SP237	stray find	fired ceramics	Irregular clay cube, rounded edges	1.49; 1.48; 1.42 (side mean 1.46)	3.91
29	VF.SP285bis	stray find	fired ceramics	Irregular clay cube, concave surfaces	1.38; 1.33; 1.24 (side mean 1.31)	3.16
30	VF14.SP284D	stray find	fired ceramics	Irregular clay cube, rounded edges, surfaces pigmented in red	1.59; 1.54; 1.52 (side mean 1.55)	5.20
31	CDA95-C45-F06	Castello di Annone cum. 45	fired ceramics	Irregular clay cube	1.60; 1.51; 1.48 (side mean 1.53)	5.38
				Dice with numbers or inscriptions		
32	VF90.1549.E32	area E, US 1549 [US 1531]	fired ceramics	Irregular clay cube, rounded edges, incised with aligned series of dot impressions	1.92; 1.70; 1.67 (side mean 1.76)	9.06
33	VF14.SP386D	stray find	fired ceramics	Irregular clay cube, numbers on the faces marked with small dot impressions	1.10; 1.09; 1.07 (side mean 1.08)	2.15
34	SP909	stray find	bone	Regular cubic dice, numbers on the faces marked with centered circles	1.16; 1.28; 1.28 (side mean 1.24)	3.62
35	CDA95-C35-F2	Castello di Annone cum. 35	fired ceramics	Irregular clay cube, two sides marked with Etruscan letters, the other four sides marked with dots and lines	1.41; 1.39; 1.37 (side mean 1.39)	3.94

Figure 2. The dice CDA95-C35-F2 from Castello di Annone with signs, dots and two Etruscan numbers in letters [12]. Scale bar: 4 mm.

Because of the puzzling presence of several cubic dice with no writing marks or numbers, Raman chemical mapping and multispectral optical imaging were also used to image the unmarked surfaces in the attempt to check for eventual phantoms indicating residues of disappeared, degraded, or canceled signs. Surprisingly, no evidence whatsoever was found of labile or disappeared inscriptions, therefore supporting the pristine unmarked state of several of the cubic dice.

3. Results of the Materials Analysis

For each find, the mean dimensions (i.e., mean side for cubes and mean diameter for spheres and circles, Table 1) were calculated for statistical analysis and comparison purposes. The details of the statistical distributions of dimensions and shapes for each typology of material are reported and discussed by Angelini et al. [11].

Here we only report the observation that independent of the size and shape of the object, the approximate density obtained by the calculated weight/volume ratio is constant, and indicates similar raw materials and manufacturing techniques. Moreover, the density of the ceramic used for the production of the dice is similar to the one of bone dice. This characteristic may be relevant in the throwing of the dice.

In each class of objects, or considering the whole set of finds, no occurrence of specific values of weight, and no system of multiple or submultiple values was observed. It is, therefore, possible to exclude the use of these objects as weights. This deduction is supported also by the absence of hooks, holes, or lateral grooves that are generally present in Bronze Age and Iron Age weights ([11], and references reported therein).

As expected from the visual observations and the preliminary density measurements, the main composing material of the ceramic bodies is a common mixture of clay and sand, fired at relatively low temperatures (600–800 °C). The range of firing temperature is deducted from the absence in the XRD analyses of clay minerals that decompose generally at about 600 °C or below, and the presence of micas that commonly start dehydroxylation transformations at temperatures of about 900 °C. Apart from the VF.SP251 washer disc that was manufactured by filing and abrading a recycled ceramic fragment of common fired pottery, all other objects were roughly put into shape by hand modeling the clay paste before firing. Most of the decorating nail marks visible on the washer-like discs (Figure 3) were also made before firing.

Figure 3. (a) nail marks visible on the flat surface of the VF-pU210-E10 washer disc; (b,c) vertical nail incisions made on the lateral edge of the VF-pd571-E41 washer disc. The scale bars are (a,b) 2 mm, and (c) 1 mm.

Indeed the systematic use of ceramics for such small pieces and for dice, in particular, is rather unusual since the great part of the investigated dice of Etruscan [13,14] or Roman [15] period are made of bone or ivory. Roman or pre-Roman dice made of amber, glass, metal, or stone are indeed known, though they are a very small fraction of those found in tombs and archaeological sites. This already indicates that the investigated assemblage from Villa del Foro has cultural aspects rather different from the main Etrusco-Roman tradition. As a matter of fact there are no known occurrences of Etruscan dice made of ceramics. The only reported ceramic dice are a few III millennium BC samples from the Middle East (Tepe Gawra, Iraq), and the Indus Valley (Mohenjo-Daro, Pakistan) [16,17], and several clay dice of the VII-VI centuries BC from Greek tombs [18]. The relevant presence of eighteen unnumbered cubic dice in the Villa del Foro context is also to be regarded as highly unusual. The only similar ceramic object known is from the nearby site of Castello di Annone (about 20 km west of Villa del Foro). Interestingly, in Castello di Annone was found also a ceramic dice with two numbers engraved in Etruscan letters (Figure 2) that were interpreted by Gambari [12] as ΘU (thu) and za (zal), respectively one and two. On the other faces of the cube are randomly impressed numerous points, and the dice seem to show the distribution: 1 (in letters)/18, 2 (in letters)/12 or 13, 9/22. The points are sometimes very close ones to the others and the decoration lines present on the edges of the faces may partially overlap the points; therefore, the readability of the points numbers seems to be not important for the use purpose of the dice. The function of these peculiar and unique dice has to be different from that of the normal dice with numbers or the unnumbered ones from Villa del Foro and Castello di Annone.

On the other hand, the use of fired ceramics for small spheres is a common occurrence in Northern and Central Italy since the Middle Bronze Age [19,20], so the presence of clay balls fits with the local use of such objects.

Concerning the colored surfaces, it should be noted that none of the washer-like discs has evidence of pigments in the surface, whereas one small sphere (VF.RS111, Figure 1b) and two dice have a visible red coating. The color is due to a hematite-rich surface layer in all cases, as unambiguously determined by XRPD and especially by the micro-Raman spectroscopy performed on a small quantity of the painted layers scrubbed from the surface of a die and a sphere (Figure 4). The other mineral phases identified in the red layer by XRD (mainly quartz, plagioclase and mica) may belong to either the ceramic body or the coloring raw material if an ochre-rich material was used. The red coating was, thus, made using crushed hematite or red ochre. In our opinion, the last hypothesis is more probable due to the availability, diffusion and large use of red ochre pigments.

One dice (VF91.1531.E38) has a visible white coating layer on the surface, which is enriched in calcite, as shown by XRPD and Raman analyses (Figure 5).

In both the red and white cases the pigmenting mineral (ochre and calcite, respectively) form uniform patinas, that appear under the microscope as a compact layer, well connected to the ceramic body and that do not show loss of powder. It was not allowed to cut a section of the colored objects, but the OM study and the absence of organic or inorganic binders (especially in the red finds) suggest that the pigments were applied to the surface on purpose, likely before firing. The observation supports the fact that the dice are completely finished, and left unmarked intentionally.

A couple of dice show irregular black areas that were shown by Raman spectroscopy to be enriched in carbon black, possibly as the result of inappropriate firing during manufacturing in reducing conditions or accidental combustion at some point in the life of the object.

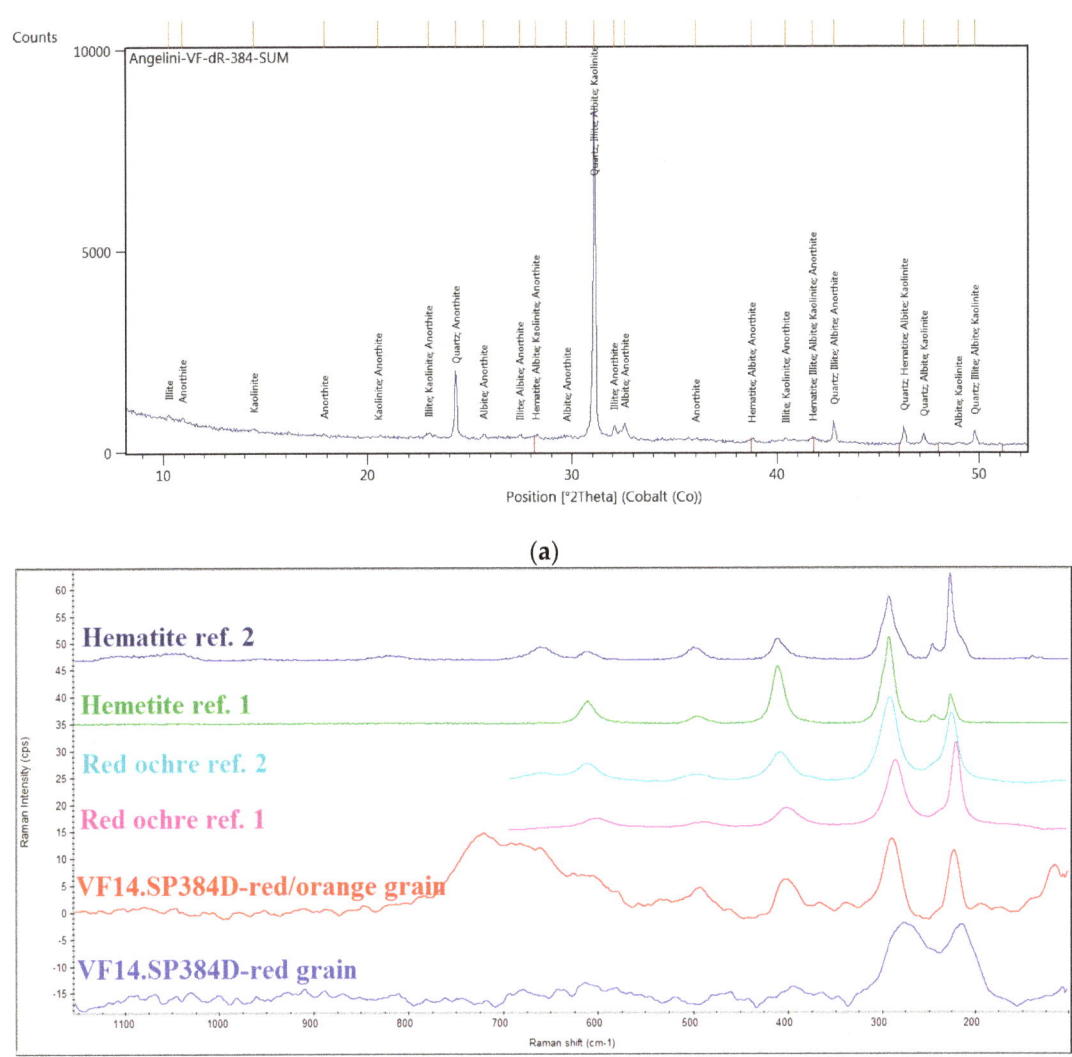

Figure 4. (**a**) XRPD diffractogram measured on the red coating of dice VF14.SP384D; (**b**) Raman spectroscopic signal of one red grain (blue spectrum) and one red-orange grain (red spectrum) present in the red coating of the same dice, compared with the spectra of two reference hematite (green and dark blue) and two red ochre samples (pink and cyan).

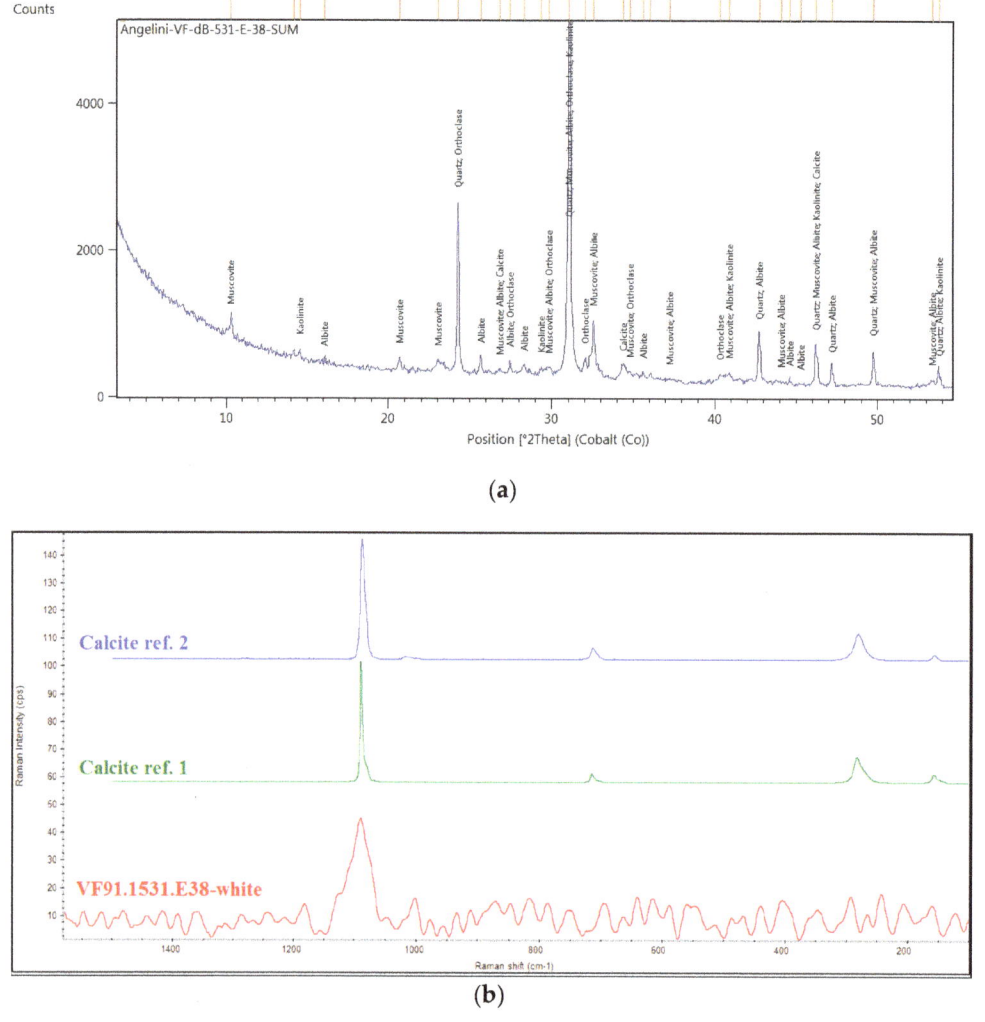

Figure 5. (a) XRPD diffractogram measured on the white coating of dice VF91.1531.E38; (b) Raman spectroscopic signal of the white coating of the same dice (red), compared with the spectra of reference calcite (green and blue).

4. Interpretation of the Objects

The occurrence of several types of small ceramics materials at Villa del Foro stimulates a number of considerations concerning their use. It is not clear whether the washer-like discs, the spheres and the dice had common functions, although their archaeological association and chronological distribution indicates that they all were in use during the lifetime of the settlement, which is considered to be an Iron Age rich trade site (*"emporium"*) along the river Tànaro [9]. During VI and V centuries BC the site was a dynamic trade site between the Ligurian populations living along the Tyrrhenian coast, the Golasecca people living in the Pavese area, and the Etruscan population of the Po Valley to the East and of Central Italy to the South. The Villa del Foro settlement was abandoned around the mid-V century BC, a period that started to witness a substantial inflow of Celtic people through the Po Valley, with subsequent intense interaction between the Etruscan and Celtic cultures [9,21]. The decline of the settlement is mainly related to the change of the trade routes towards the Western Alps (e.g., the way along the Scrivia valley in the direction of Milan and Como increase its importance in time) [9].

Several proposals have been put forward to interpret the use of ceramic washers and balls, such as cultural function, seal for containers of different types, loom weights, pieces for board games or kids' toys, tokens for computation support for pottery. A detailed discussion of the evidence in favor or against each one of the hypothetical functions can be followed in Paltineri [10] and Angelini and Bettineschi [11]. Here we just recall that based on the archaeological context [10], the use/wear traces, and the technological properties and weight of the objects, the most likely function should be either as computational tools, or pieces for board games, likely associated to the dice. The interpretation as tools for calculations is stimulating, and it could in principle be linked to the vast trading activities of the site. Further, it is known that the earliest calculation tools in the Middle East were composed of tokens having different geometrical forms [19]. However, the association of the washers and spheres with dice, especially with the unusually large number of un-inscribed cubic dice, may well indicate an assemblage of board pieces with different shapes and colors, given the observed pigment coating. Indeed the lack of marks or inscriptions in all three types of materials supports this interpretation, rather than counting or divinatory activities.

On the other hand, there is little doubt that the dice with incised numbers were used for games. The dice having stick, parallelepiped, or cubic form were in use in the Indus Valley, Mesopotamia, and Egypt at least from the III millennium BC [16,17]. The traditional interpretation is that the dice games were introduced to Central and Northern Italy by Etruscans through the Greek world. There is no evidence of dice in Central Europe and in the Celtic or pre-Celtic cultures West of Italy [16,22,23] before Roman times, and this sets the wider cultural context of the Villa del Foro site during the VI and V centuries BC: a trading place between different cultures. Possibly one of the places where dice games were introduced by Etruscans to nearby populations of North-Western Italy.

There are two dice with numbers at Villa del Foro: (1) sample SP909 (Figure 6), made of bone, with incised numbers in the 1–6 2–5 3–4 configuration, that is the opposite faces sum to 7 (also defined as the "*sum 7 rule*" [14,15]), and (2) sample VF14.SP386D (Figure 7), made of ceramics, with impressed small points indicating numbers in the 1–6 2–4 3–5 configuration. The detailed analysis of the distribution of the numbers on the sides of a dice is a very interesting operation [11] and it cannot be fully detailed here. Based on previous work [11,13–15,24–26] we recall some of the main issues involved:

(a) disregarding the rotation of the dice and the relative disposition of the points in the side of the cube, there are only 15 possible coupled configurations of numbers on opposite faces [13–17,25];

(b) the two most frequent configurations in ancient dice are the 1–6 2–5 3–4 configuration ("*sum 7 rule*") that is virtually the universal configuration found in modern dice, and the 1–2 3–4 5–6 configuration ("*difference 1 rule*"). It turns out that none of the two numerical distributions are naturally inserted on the face of the cube without cultural imprinting [24]. This is strong proof of the social and cultural transmission values of dice games [2,3];

(c) in ancient Italy, the dice were introduced by Etruscans, the early dice had a parallelepiped shape, then evolved into the cubic one. The early dice (VII-VI century BC) on both sides of the Apennines (i.e., Etruria, South of the Apennines, and Etruscan Po Valley) were mostly of parallelepiped shape and have a neat prevalence of the "*difference 1*" configuration", then largely shifting to the prevalent cubic form with the "*sum 7*" configuration in the V century BC. In Bologna (*Felsina*), the capital of Etruria in the Po Valley, the old shape and numerical configuration seems to maintain prevalence throughout the V century BC. Starting from the IV century BC the latter configuration is virtually the only one present and it continued to be prevalent in the Roman world [9,11–13]. Because of the conservative character of dice in Etruscan Bologna, the bone dice of Villa del Foro (SP909, Figure 6) seem to testify to contact with Etruscans from the Tyrrenian area, rather than the Po Valley.

(d) in all periods, there is a small fraction of dice having a numerical configuration different from the two prevalent ones. In most cases, we have a doubling of a number and a missing one (for example 1-1 3-4 5-6), and sometimes we observe the doubling of three numbers and the omission of the remaining ones (for example 1-1 5-5 6-6 or 2-2 3-3 4-4). Such dice were sometimes interpreted as manufacturing mistakes or intentional variations for cheating, though the most probable use is in connection with board games [11]. Interestingly, many of these anomalously numbered dice were found in association with board pieces, such as glass checkers [11,16,27].

Within this context, the ceramic dice from Villa del Foro (VF14.SP386D, Figure 7) has both a numerical configuration (1–6 2–4 3–5) and the nature of the material (ceramics) that has never been observed in Etruscan dice [11]. Once more, such features confirm the use of the Villa del Foro dice at the boundaries of Etruscan cultural influence.

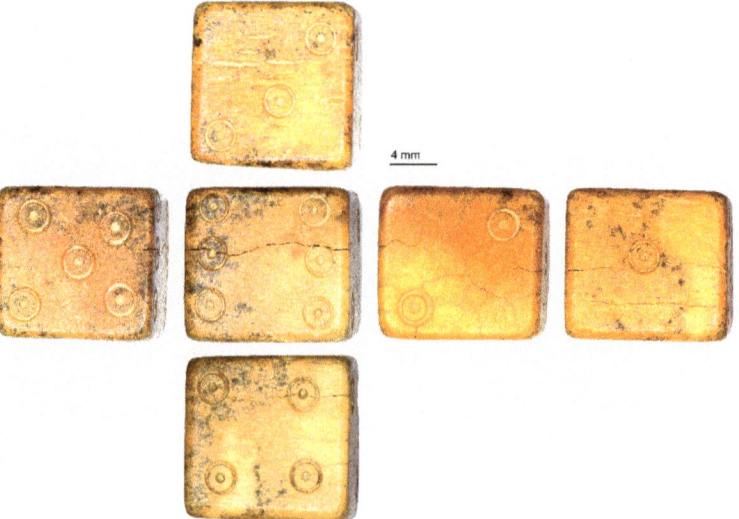

Figure 6. The dice SP909 from Villa del Foro, with numbers 1–6 incised in the "*7 rule*" configuration. Scale bar: 4 mm.

Figure 7. The dice VF14.SP386D 9 from Villa del Foro, with numbers 1–6 incised in the 1-6 2-4 3-5 configuration. Scale bar: 4 mm.

5. Conclusions

The very unusual dice found in Castello d'Annone, bearing Etruscan letters and randomly distributed dots (Figure 2), is a *unicum* that may be somehow compared to the well-known Etruscan dice with numbers in letters from Toscanella, discussed by various authors and interpreted as being used for ritual or divinatory purposes [28,29]. The same function has been suggested for the inscribed dice from Castello di Annone [12]. The features of these dice reflect contact with the Etruscan culture, although it is made of ceramic, which is not employed in the making of Etruscan dice. The presence of unnumbered ceramic dice in the site, and the peculiarity of the marks suggest a local production. The ceramic dice from Villa del Foro with irregular numbers of points, that are not well readable on three faces (Figure 1d), differs from all the other dice found in the settlement, and possibly it was used for similar ritual or divinatory purposes as the Castello di Annone dice bearing letters and signs.

Overall, the materials analysis of the small ceramic objects excavated from the Villa del Foro archaeological site confirms their exceptional importance both for the large quantity of recovered finds and for the reconstruction of the social and cultural context of this "emporium". Based on the measured data and the archaeological information, it is proposed that the ceramic pieces, in the form of washers, spheres, and uninscribed cubes, were used as counters in board games. The use of these objects as tokens for computation cannot be excluded, although in our opinion it is considered less probable for several reasons. At first, the recovery of two dice with numbers from 1–6 testify to the presence of ludic activities in the site. Moreover, in the Etruscan world game counters were widespread and often found in association with dice, as discussed below. The connection of Villa del Foro with the Etruscan culture is proved by different finds recovered in the site [9] and by the presence of cubic dice with numbering 1–6 and a configuration in agreement with the "*sum 7 rule*". It is therefore probable that the numerous ceramic washers, cubes and spheres had the same function as the Etruscan counters made with different materials. The red or white color intentionally given to some sphere and cubes from Villa del Foro support this idea.

No remains of the board were found in Villa del Foro or in other sites in Northern Italy or in the Etruscan territories; so it is impossible to know what kind of game was played. On the other hand, the use of board games is well proven by the presence of numerous counters made with different colors and materials (often associated with dice), found in Etruscan tombs [10,11,19,22,27]. Based on the iconographic study of ceramic decorations and on a few engraving in stone found mainly in Greek and Roman contests, many studies and hypotheses of the types of board game used have been published (e.g., [1,3,18,21]). Likely, the same or similar types of board games were played in Villa del Foro.

Author Contributions: Conceptualization, I.A., C.B., M.V. and G.A.; methodology, I.A., C.B.; formal analysis, I.A., C.B.; investigation, I.A., C.B.; resources, I.A., G.A.; data curation, I.A.; writing—original draft preparation, G.A.; writing—review and editing, G.A., I.A., C.B., M.V. All authors have read and agreed to the published version of the manuscript.

Funding: This research received no external funding.

Institutional Review Board Statement: Not applicable.

Informed Consent Statement: Not applicable.

Data Availability Statement: Not applicable.

Acknowledgments: Rita Deiana (Università di Padova) is acknowledged for help in the measurement with multispectral images. We thank the four anonymous referees who improved the quality of the paper.

Conflicts of Interest: The authors declare no conflict of interest.

References

1. Finkel, I.L. *Ancient Board Games in Perspective: Papers from the 1990 British Museum Colloquium with Additional Contributions*; British Museum Press: London, UK, 2007.
2. de Voogt, A.; Dunn-Vaturi, A.E.; Eerkens, J.W. Cultural transmission in the ancient Near East: Twenty squares and fifty-eight holes. *J. Archaeol. Sci.* **2013**, *40*, 1715–1730. [CrossRef]
3. Hall, M. Whose Game is it Anyway? Board and Dice Games as an Example of Cultural Transfer and Hybridity. *Archéol. Hist. Ancienne* **2019**, *6*, 199–212. [CrossRef]
4. *Art du jeu, jeu dans l'Art de Babylone à l'Occident Medieval*; Exhibition Catalog, 28 November 2012–4 March 2013; Musée de Cluny, Réunion des musées Nationaux–Grand Palais: Paris, France, 2012.
5. *Jouer dans L'antiquité*; Musée d'Archéologie Méditerranéenne-Centre de la Vieille Charité, Exhibition Catalog, 22 November 1991–16 February 1992; Musées de Marseille-Réunion des Musées Nationaux: Marseille, France, 1991.
6. Lambrugo, C.; Della Torre, C. *Il Gioco ed i Giochi nel Mondo Antico: Tra Cultura Materiale ed Immateriale*; Edipuglia: Bari, Italy, 2013.
7. Dasen, V. (Ed.) *Ludique*; Catalog of the Exhibition Ludique! Jouer dans l'Antiquité, 20 June–1 December 2019. Lugdunum-Musée et Théâtres Romains, Gent, Snoeck: Gent, Belgium, 2019. Available online: https://doc.rero.ch/record/330615/files/dasen_2019_ludique.pdf (accessed on 30 October 2021).
8. Dasen, V. (Ed.) Dossier Locus Ludi: Les dés atypiques. In *Instrumentum: Bulletin du Groupe de Travail Européen sur L'artisanat et les Productions Manufacturées dans L'antiquité*; 2020; Volume 52, pp. 26–46. Available online: https://zenodo.org/record/5105906#.YgQoP_zSLIU (accessed on 30 October 2021).
9. Venturino, M.; Giaretti, M. (Eds.) *Villa del Foro. Un Emporio Ligure tra Etruschi e Celti, (ArcheologiaPiemonte, 8)*; De Ferrari: Genova, Italy, 2021.
10. Paltineri, S. I fittili non vascolari. Rondelle, sferette, dadi e il problema degli elementi da gioco. In *Villa del Foro. Un emporio ligure tra Etruschi e Celti, (ArcheologiaPiemonte, 8)*; Venturino, M., Giaretti, M., Eds.; De Ferrari: Genova, Italy, 2021; pp. 413–426.
11. Angelini, I.; Bettineschi, C. Gli elementi da gioco. Analisi archeometriche e numerali. In *Villa del Foro. Un Emporio Ligure tra Etruschi e Celti, (ArcheologiaPiemonte, 8)*; Venturino, M., Giaretti, M., Eds.; De Ferrari: Genova, Italy, 2021; pp. 427–444.
12. Gambari, F.M. Il dado fittile con lettere etrusche. In *La Memoria del Passato. Castello di Annone tra Archeologia e Storia (ArcheologiaPiemonte, 2)*; Venturino Gambari, M., Ed.; Linelab Edizioni: Alessandria, Italy, 2014; pp. 273–274.
13. Nociti, V. *Dadi e tessere dall'Etruria (Tarquinia, Orvieto, Chiusi), Tesi di Laurea Magistrale*; Università degli Studi di Milano: Milano, Italy, 2007.
14. Artioli, G.; Nociti, V.; Angelini, I. Gambling with Etruscan dice: A tale of numbers and letters. *Archaeometry* **2011**, *53*, 1031–1043. [CrossRef]
15. Eerkens, J.W.; de Voogt, A. The evolution of cubic dice: From the Roman through Post-Medieval period in the Netherlands. *Acta Archaeol.* **2017**, *88*, 163–173. [CrossRef]
16. Küchelmann, H.C. Why 7? Rules and exceptions in the numbering of dice. *Palaeohistoria* **2018**, *59*, 109–134. [CrossRef]

17. Ineichen, R. *Würfel und Wahrscheinlichkeit–Stochastisches Denken in der Antike*; Spektrum Akademischer Verlag: Heidelberg, Germany, 1996.
18. Ignatiadou, D. Luxury Board Games for the Northern Greek Elite. *Archéol. Hist. Ancienne* **2019**, *6*, 144–159. [CrossRef]
19. Marchi, E.; Pancaldi, P.; Tesini, M. Palline. Possibile valenza magico-rituale delle sferette fittili tra Bronzo Recente e Bronzo Finale nella Pianura Padana. In *Pagani e Cristiani. Forme e Attestazioni di Religiosità del Mondo Antico in Emilia*; All'Insegna del Giglio: Firenze, Italy, 2013.
20. Pizzi, C. *L'abitato dell'età del Bronzo di S. Caterina Tredossi (Cremona). I materiali Conservati Presso i Musei di Milano e Cremona*; Comune di Milano-Raccolte Archeologiche e Numismatiche: Milano, Italy, 2006.
21. Schädler, U. Greeks, Etruscans, and Celts at play. *Archéol. Hist. Ancienne* **2019**, *6*, 160–174. [CrossRef]
22. de Chavagnac, L. Dés Pré Romains. In *Ludique*; Dasen, V., Ed.; Catalog of the Exhibition Ludique! Jouer dans l'Antiquité, 20 June–1 December 2019; Lugdunum-Musée et Théâtres Romains, Gent, Snoeck: Gent, Belgium, 2019; pp. 110–111.
23. Blasco Martín, M. Dados y fichas de la Edad del Hierro en la Península Ibérica. *Arch. Prehist. Levant.* **2016**, *XXXI*, 241–260.
24. De Voogt, A.; Eerkens, J.W.; Sherman-Presser, R. Production bias in cultural evolution: An examination of cubic dice variation in experimental and archaeological contexts. *J. Anthropol. Archaeol.* **2015**, *40*, 151–159. [CrossRef]
25. Poplin, F. Numération et orientation des dés antiques et médiévaux. *Instumentum* **2012**, *36*, 30–34. [CrossRef]
26. De Voogt, A.; Eerkens, J.W. Cubic Dice: Archaeological Material for Understanding Historical Processes. *Kentron. Rev. Pluridiscip. Monde Antiq.* **2018**, *34*, 99–108. [CrossRef]
27. Macellari, R. *Il Sepolcreto Etrusco nel Terreno Arnoaldi di Bologna (550-350 a.C.)*; Comune di Bologna: Bologna, Italy, 2002.
28. Savelli, A. I dadi del Museo Civico di Bologna e il problema dei numerali etruschi. *Strenna Stor. Bolognese* **1976**, *XXVI*, 271–290.
29. Agostiniani, L. Sui numerali etruschi e la loro rappresentazione grafica. *AION* **1995**, *17*, 21–65.

Article

Materials Inspiring Methodology: Reflecting on the Potential of Transdisciplinary Approaches to the Study of Archaeological Glass

Sara Fiorentino *, Tania Chinni and Mariangela Vandini

Department of Cultural Heritage, Ravenna Campus, University of Bologna, 40126 Bologna, Italy; tania.chinni2@unibo.it (T.C.); mariangela.vandini@unibo.it (M.V.)
* Correspondence: sara.fiorentino2@unibo.it

Abstract: The contribution that materials science has made to the understanding of ancient glass-making is unquestionable, as research undertaken in recent decades has extensively demonstrated. Archaeological glass is far from being a homogeneous class of materials, encompassing objects made for different uses, manufactured in different periods and geographic areas, with a variety of tools and working techniques. If all these factors are not adequately considered when approaching the study of ancient glasses from an archaeometric perspective, data obtained by analyses can incur the risk of being less informative or even misinterpreted. Moving from previously performed research, this paper is aimed at reflecting on the potential of synergistic approaches for the study of archeological glasses, based on the interrelation among different disciplines and fostering the integration of archaeological and historical knowledge with data-driven scientific analyses.

Keywords: archaeological glass; archaeometry; VIS–RS; SEM–EDS; Raman spectroscopy; XRPD; EPMA; LA–ICP–MS

1. Introduction

The study of objects belonging to our material culture cannot ignore a persevering and constructive dialogue between different fields of knowledge. Only a well-integrated, transdisciplinary approach can pave the way for an in-depth understanding of historical, cultural, socioeconomic, and technological issues for which the objects of our past are silent witnesses.

Although in the last few decades there has been an increasing tendency towards combined archaeological and archaeometric studies [1,2], this has not always resulted in long-lasting cooperation among disciplines. On the one hand, archaeology often turns to archeometry to answer open questions mainly related to the provenance and dating of the objects or as a scientific support to confirm previously formulated hypotheses. On the other hand, archaeometric studies leave a space that is not always adequate for archaeological and/or historiographic considerations, sometimes leading to a discussion of extreme detail of a single analytical data which, however, is detached from a broader contextualisation. The above considerations are also valid for archaeological glass, a category whose applied research is still affected by a subdivision of roles based on the specific knowledge in a particular field as well as by a lack of systematic integration of data across disciplines. All this to the detriment of a real joint research action, which is fundamental for understanding objects from the past beyond their material components and as vectors of culture. More specifically, though a noticeable broadening of the glass research community has occurred in the last decades, the need to move from a traditional data gathering to a more interpretative phase has been highlighted [3]. Research should guide toward a better understanding of the reasons and practices behind the compositional and technical developments of the ancient glass industry, in the attempt to deepen the current understanding of glass production, trade, and technologies and to contextualise this specific material within the sociocultural and economic framework of past societies. The discernment of such a multifaceted scenario cannot ignore a synergistic and complementary approach between different disciplines. A sound integration of analytical data into archaeological and, by extension, anthropological research, is the keystone that can lead to a more comprehensive understanding of ancient glass industries and related technological aspects.

The establishment of an enduring transdisciplinary approach could overcome several critical issues affecting applied research on material culture, with specific reference to glass production. In the first place, one should be cautious in comparing archaeometric data relating to different categories of objects (i.e., tableware with mosaic tesserae and/or ornaments), as sociocultural and economic reasons behind the manufacture of these objects could lead to erroneous conclusions. Comparing objects destined for different uses, albeit coming from the same context, can mislead the interpretation of archaeometric data and lead to hypotheses that are not sustainable from an archaeological perspective. Hence, there is a need to overcome the traditional method of comparison, mainly based on associating data pertinent to types of glass objects very different from each other. Another critical issue to pay particular attention to is the number of finds to be analysed: obtaining conspicuous datasets should not be the key criterion when selecting the finds. If the analysed fragments are not "diagnostic", meaning they are not attributable to distinctive typologies with verified chronology, the achieved dataset, though numerically consistent, might not provide useful information. Prior to the analytical phase, it is of primary importance to understand the context where the finds were unearthed, as well as the related sociocultural aspects.

As a consequence, an actual co-participatory approach between disciplines related to different research fields is the only tool capable of overcoming the limits that, by its intrinsic nature, each discipline considered individually has. Without prejudice to the need for everyone to work in their own specific field of study, it is necessary to overcome the still feebly implemented reading and synergistic discussion of the results that emerge from the various disciplinary areas involved.

Through an examination of selected previously studied assemblages from different types of contexts, this paper aims to reflect on the potentialities stemming from a transdisciplinary approach applied to the study of archaeological glass. Whatever the research field, ancient glasses are strongly heterogeneous materials: from a historical-archaeological perspective, glass assemblages from excavated contexts comprise different objects intended for different uses (i.e., glassware, ornaments, and architectural decorations), often made in different places and workshops, using different tools and techniques. Should these elements not be properly taken into consideration, data obtained by archaeometric analyses could incur the risk of being uninformative. Framed in a broader context and without any intent to criticise previous research, this paper fosters the promotion of a synergistic approach in the study of archaeological glass, setting the basis for an enduring integration of different disciplines.

2. Materials and Methods

Previously studied assemblages of archaeological glass available at the Conservation Science Laboratory of the Department of Cultural Heritage (University of Bologna-Ravenna Campus) have been selected for this study: 13th–14th century ampoules, nuppenbecher and kropfflasche from the monastic complex of San Severo (Classe, Ravenna, Italy) [4]; 14th–16th century gambassini drinking vessels from the monastic complex of San Severo (Classe, Ravenna, Italy) and from the Rontana Castel (Brisighella, Ravenna, Italy) [5]; 8th century mosaic glass tesserae from the qasr of Khirbet al-Mafjar (Jericho, Palestine) [6]; and 8th century mosaic glass tesserae from the Great Mosque of Damascus (Syria) [7,8]. When other assemblages of materials from previous studies which have not been carried out by the authors are taken as examples, references are provided in the text.

An Olympus S761 stereomicroscope (magnification up to 45×, Olympus Corporation, Shinjuku, Tokyo, Japan) associated with an Olympus Soft Imaging Solutions GMBH model

SC100 camera (Olympus Corporation, Shinjuku, Tokyo, Japan) was used for preliminary morphological observations and documentation.

For Visible Reflectance Spectrometry (VIS–RS), a MINOLTA CM-2600d portable spectrometer was used. The system is equipped with an internal integrating sphere of 56 mm diameter, in reflectance geometry d:8°, with three Xenon pulsed lamps, and a D65 illuminant was used; calibration was performed against a $BaSO_4$ standard plate; the spectral range is 400–700 nm, with a spectral resolution of 10 nm and the area of sight of 3 mm diameter. Spectra Magic software was employed to elaborate data; specular component excluded (SCE/0) was selected, according to the literature [9].

Microsamples were taken from either objects or diagnostic fragments for performing micromorphological, microtextural, and compositional analyses. Sections were prepared by embedding microsamples in a polyester resin then polished by using abrasive papers of decreasing grain in order to expose the embedded samples avoiding any losses.

A Scanning Electron Microscope (SEM) back-scattered electron signal (BSE) was used for the inspection of the morphological features of the inclusions coupled with Energy Dispersive Spectroscopy (EDS) spot measurements to achieve a preliminary qualitative and semiquantitative elemental analysis of the inclusions themselves. Analyses were carried out on polished and carbon-coated sections. Images and EDS spectra were collected on carbon-coated crosssections, using an ESEM FEI Quanta 200, equipped with an EDAX energy dispersive spectrometer. Analyses were performed in high-vacuum using an acceleration voltage of 25 kV and an energy resolution of ~200 eV; working distance was set at 10 mm, spot size was between four and five µm.

Micro-Raman spectra were collected by using a Bruker Senterra dispersive Raman spectrometer equipped with an integrated Olympus BX40 microscope. A 785 nm He–Ne laser was employed, in the 300–3500 cm^{-1} region. Analytical measurements were performed with a 50× long working distance objective, operating at a power of 10 mW (red and blue samples) or 25 mW (yellow and green samples) with a spectral resolution of 3.5 cm^{-1}. Raman measurements were performed on polished sections after carbon-coating removal.

X-ray Powdered Diffractometry (XRPD) analyses were performed on finely powdered samples manually pressed on a Ag sample holder in a Rigaku Miniflex diffractometer employing $CuK\alpha_1$ radiation in the range 2θ: 4°–64°, θ scan speed: 1°min^{-1}.

To determine the bulk chemistry of all samples under study, Electron Probe Microanalysis (EPMA) was performed on polished and carbon-coated sections. Chemical analyses of major and minor elements (Na, Mg, Al, Si, P, S, Cl, K, Ca, Ti, Cr, Mn, Fe, Co, Cu, Sn, Sb, and Pb) were performed using a CAMECA–CAMEBAX equipped with four scanning Wavelength Dispersive Spectrometers (WDS). A beam current of 2 nA, an acceleration voltage of 20 kV, and a spot size of 5 µm were used for Na, K, Si, and Al; for all other elements, a beam current of 20 nA, an acceleration voltage of 20 kV, and a spot size of 1 µm were used. Synthetic pure oxides were used as standards for Al, Cr, Fe, and Sn, synthetic $MnTiO_3$ for Mn and Ti, wollastonite for Si and Ca, albite for Na, periclase for Mg, PbS for Pb, orthoclase for K, apatite for P, sphalerite for S, Sb_2S for Sb, and pure elements for Co, Cu, and Ni. SMITHSONIAN GLASS A standard [10] was also employed as a reference sample. Ten points were analysed on each sample, and the mean values were calculated. The measured accuracy for the analysed elements was better than 3%. The standard deviations among the analysed points resulted in between 1% and 3% and 3% and 5% for major and minor constituents, respectively. The detection limit for the minor elements was between 0.01 wt% and 0.04 wt%. The correction program was based on the PAP method [11] and was used to process the results for matrix effects.

Laser Ablation–Inductively Plasma Coupled–Mass Spectrometry (LA–ICP–MS) was carried out to determine the concentration of 37 trace elements. Analyses were performed by a Thermo Fisher X-Series II quadrupole based ICP–MS coupled with a New Wave ablation system with a frequency quintupled (λ = 213 nm) Nd:YAG laser. Laser repetition rate and laser energy density on the sample surface were fixed at 20 Hz and ~18 J/cm^2, respectively. Analyses were carried out using a laser spot diameter of 100 µm on the same polished samples used for EPMA after carbon-coating removal. Due to the highly heterogeneous microstructure of the tesserae, six points were analysed on each sample and the mean values were then calculated. External calibration was performed using NIST 610 and 614 glass as external standards; NIST 612 was also used as a secondary reference sample to check precision and accuracy (Pearce et al. 1997). ^{29}Si was employed as an internal standard, whose concentration was determined by EPMA following the method proposed by Longerich and colleagues [12]. The distribution of REE and of the other trace elements was analysed by normalising the data to the upper continental crust [13].

3. Results and Discussion

3.1. Understanding Archaeological Context: The First Step toward a Tailored Selection of Materials

The most recurring questions related to the study of archaeological materials are aimed at understanding whether a local production can be identified. This issue is of a particular relevance for glass, where the very sporadic occurrence of furnace remains makes it extremely challenging to map the production sites and distinguish them from the working sites. If the context where materials have been unearthed shows evidence of a production/working activity of glass (i.e., furnace remains, production waste, and tools), the selection of the assemblage to be investigated is generally aimed to characterise all these indicators. Therefore, the choice to carry out archaeometric analyses on materials that can be typologically heterogeneous (such as production indicators, scraps, and finished objects) is justified by the need to identify raw materials used in glassmaking. An example is, in this regard, the ancient port of Classe (Ravenna, Italy), one of the most important trade centres between the 5th and 8th centuries CE in the Northern Adriatic area. Archaeological excavations identified the main context for glassworking inside one of the warehouses built at the beginning of the 5th century CE, where a small circular kiln had been unearthed; around the kiln, a massive amount of glass fragments and glassworking waste (973 finds) was brought to light in 2001 [14]. Results from archaeometric analyses demonstrated that, in the 5th century CE, the secondary glass workshop in Classe was dedicated to the shaping of vessels starting from raw glass chunks and, possibly, cullets. Evidence of glass remelting and glassblowing were found, while no evidence of primary production was uncovered [15]. Moreover, comparisons between Classe and Aquileia, the two most important Late Antique archaeological sites of Northeastern Italy, were undertaken to shed light on the role of the Northern Adriatic area in glass production, trade, and consumption. The comparison showed that, unlike Classe, correlations did not exist in Aquileia between specific types of objects and the compositions of glass; this data has further strengthened the hypothesis of a local shaping of specific objects at Classe [16].

An analogous approach in the selection of materials to be studied can be applied to cargos from shipwrecks. The case of the wreck of the Iulia Felix ship, which sank in the 2nd century CE off the coast of Grado (Friuli-Venezia Giulia, Italy) is, in this sense, thorough. The archaeological and archaeometric study of the assemblage of glass materials was performed on cullets and finished objects to achieve data on colourless and coloured Roman glass production technology. The strong evidence of compositional variability among the Iulia Felix assemblage strengthened the hypothesis of a dispersed production model for Roman glassware and the common practice of recycling in Roman ages, especially for daily-use vessels [17,18].

If archaeological glass is found at multilayered contexts and/or where there is no univocal evidence of local production, the approach to the selection of materials to be studied needs to be different. Multilayered contexts are particularly challenging to be approached: inhabited for more or less extended periods, they generally are urban, monastic, and/or residential sites, where structures that can be interpreted as small kilns, though without sound evidence of glass production (i.e., no production/processing indicators recovered), can also be found. When these contexts are faced, an archaeometric study of numerically consistent sets of fragments (without paying due attention to the multilayered structure of the site and often comparing data obtained from the analysis of fragments pertaining to different objects and chronological frames) can rarely result in a meaningful and exhaustive scenario. The result is often a multiplicity of hypotheses regarding the import of objects from other sites and, when no comparisons can be found with other sites, the risk is to hypothesise local productions without sound archaeological evidence.

Recent studies have highlighted the potential associated with adopting a synergistic transdisciplinary approach when working on multilayered contexts. The starting point is represented by an in-depth tailored archaeological study of the assemblage functional to identify distinctive shapes and their chronology. The results obtained by chronotypological study of the objects can, in fact, outline different scenarios to assess; the most common are those characterised by a) specific groups of objects with the same chronology; (b) a greater statistical incidence of a specific distinctive shape (or a few of them). Depending on whether the chronotypological study of the finds identifies one or the other scenario, key issues to which the archaeometric analyses will be called to contribute will be evaluated. An exhaustive example of the approach centred on the study of specific types of objects falling within the same chronological frame is provided by Sedlackova and colleagues [19]. The study aimed to investigate the production and import of glass objects in Bratislava involving finds of tableware and window glass. The preliminary archaeological study allowed the attribution of the finds to three main timeframes verified by the occurrence of distinctive shapes and decorative features: 13th, late 13th–14th and mid-15th century CE. The subsequent archaeometric investigations made it possible, for each of the examined periods, to ascertain the composition of the glass and to establish comparisons with chronologically compatible Central Europe productions shedding light on commercial and political orientations. The study is an example of how informative data can also be obtained by working on a small number of fragments, as long as well contextualised from a chronotypological perspective to set up comparison with the same types of objects found in other geographical areas.

The same methodological approach has been applied to a first selection of materials from the monastic complex of San Severo in Classe (Ravenna, Italy) [4]. The chronotypological study of the material identified the occurrence of three particular types of glass objects, although not present in statistically high numbers: ampoules, nuppenbecher, and kropfflasche, dated to the 13th–16th century CE. The close similarities, in terms of chemical features, between these objects from San Severo and comparable finds from Czech Republic, further strengthened the hypothesis of commercial contact between Central Europe and Italy in the Middle Ages. The scope of both studies, focused on a few distinctive shapes, was to trace, through an interrelation between chronotypological and archaeometric study, contacts between two geographically distant areas, united by the occurrence of the same types of glass objects. The monastic complex of San Severo can be taken as an example of another approach applicable to archaeological glass, based on the study of one (or a few) statistically relevant shape(s). This is the case of the daily-use drinking glass beaker known as gambassino [5]. Recently published research aimed at deepening the historical, archaeological, and compositional knowledge between the variants of this object, found in two contemporary sites located at a distance of about 50 km from each other: the monastery of San Severo and the Rontana castle (Brisighella, Ravenna, Italy). For both sites, the gambassino was the shape with the greatest statistical impact, accounting for over 70% of the entire assemblage of unearthed glass materials [14]. Moving from a local scale, the study then expanded to regional and transregional comparisons involving central-northern Italy and the Balkan side of the Adriatic Sea. Thanks to an interrelation among archaeological, historical, and archaeometric data, evidence has been underpinned for the movement of finished objects, raw materials, and artisans in the considered timeframe: the displacement of artisans and the circulation of raw materials, the recurrence of this beaker in numerous excavation contexts in the northern Adriatic area, as well as the identification of compositional groups different from each other but overlapping among assemblages, has led to the hypothesis of a "widespread production" model of the common drinking beaker known as gambassino.

The aforementioned studies stand as examples of the great variety of approaches applicable to the study of archaeological glass. The selection of one or the other approach cannot be separated from the in-depth knowledge of the archaeological context, paying particular attention to the type of settlement (i.e., urban, rural, residential, or monastic) and to its continuity of life across time. Being aware of the context is, thus, the first step to be taken. The next step is the in-depth study of the finds, of the way they are linked to the chronology of the site, and of their possible connections with other contexts. All these elements contribute to the definition of the questions underlying the selection of the samples to undergo archaeometric analyses, so that they can provide suitable data to support hypotheses of production and/or provenance of the artefacts under study.

3.2. Archaeometric Analyses: Defining the Analytical Approach Based on Material Features

Once the selection of materials to be analysed has been completed, the setup of the analytical protocol may vary according to the features of the materials themselves. An in-depth discussion of the potential and limits of all possible analytical techniques applicable to the study of archaeological glass would be beyond the scope of this article. Therefore, in the following discussion, reference will be made to the most commonly used analytical techniques for archaeometric studies on glass to foster data comparability.

The first criterion influencing the choice of analytical protocol is whether the materials to be analysed are naturally or intentionally coloured glasses. The archaeometric approach to the study of nonintentionally coloured glass (which means without the intentional addition of colouring and opacifying agents to the base glass) is simpler than deeply coloured (and often opaque) glass. The latter is, in fact, characterized by greater heterogeneity in terms of microstructure and compositional features, due to the addition of specific raw materials aimed at imparting the desired colour and degree of opacity. The setup of a more articulated analytical approach is needed based on the interrelation of different investigation techniques to achieve an in-depth characterization of colouring and opacifying agents. When dealing with naturally coloured glass, the archaeometric approach is based on the use of fewer analytical techniques, the primary objective being the determination of the chemical composition of the base glass and, if needed, the provenance of the materials used as vitrifying agents. Figures 1 and 2 show block diagrams of the two different analytical approaches applicable to the study of naturally and intentionally coloured glass; it is always advisable to carry out an observation and documentation of the samples in optical microscopy, preliminary to the filing of the fragments/objects under study.

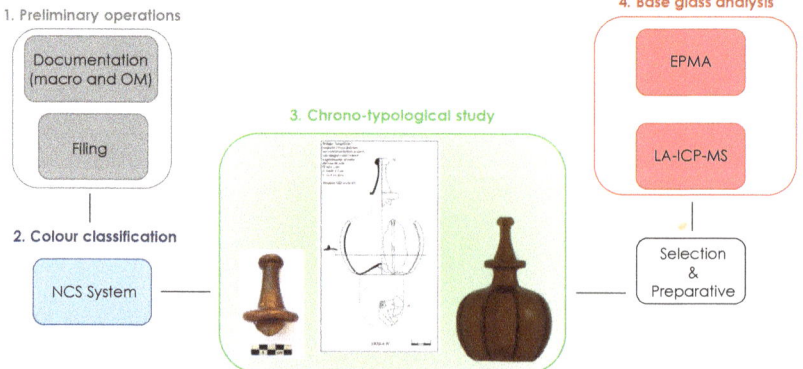

Figure 1. Block diagram of the proposed analytical approach for naturally coloured archaeological glass.

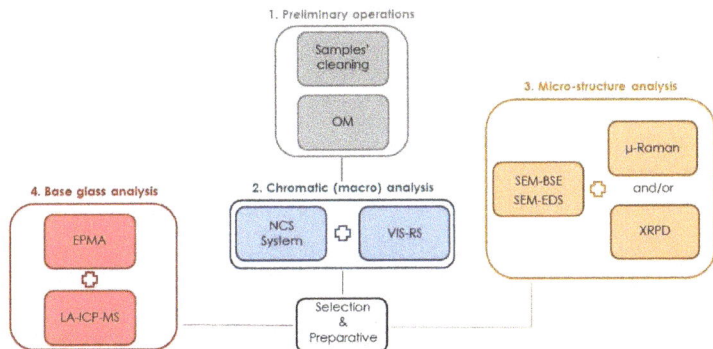

Figure 2. Block diagram of the proposed analytical approach for coloured archaeological glass.

When archaeometric analyses have to be performed on intentionally coloured glasses, it is highly recommended to start with an in-depth study of the colouring (and opacifying) phases. As more extensively discussed elsewhere [8], since deeply coloured glasses show highly heterogeneous microstructures and microtextures, a thorough characterisation of colouring and opacifying phases should be carried out before the bulk composition of the vitreous matrix to avoid misunderstandings in data processing and interpretation.

A combined Standard Colour System Chart (such as NCS Index or PANTONE) and Visible Reflectance Spectroscopy (VIS–RS) approach is here proposed as a starting point. It can support, in fact, an objective definition of the chromatic hues and shades of deeply coloured glasses avoiding any subjective nomenclature. The example par excellence is represented, in this case, by a particular category of glass-based artefacts: mosaic tesserae. For these small cubes of coloured glass, a chronotypological study is unfeasible; therefore, colours and opacity are the only macrofeatures that can be used to methodically select among copious assemblages of samples to be studied. NCS coordinates can be used to effectively separate the tesserae into preliminary chromatic macrocategories (i.e., yellow, green, blue, red, and black). These groups can be defined based on NCS coordinates and, more precisely, by taking the second part of the NCS-notation into account, which describes the hue by means of a numerical code. For example, a tessera with NCS-notation S 2030-G70Y will be described as of a yellow colour and code G70Y indicating a colour shade described as a yellow (Y) with 70% resemblance to yellow and a 30% resemblance to green (G). It is possible, in this way, to avoid the use of definitions such as "greenish yellow" or "yellowish green", eliminating a first important degree of subjectivity in the definition of colour. For NCS evaluations to be accurate, it is important that observations are made in controlled repeatable daylight conditions avoiding any artificial source of light.

After preliminary NCS-aided discrimination between chromatic macrocategories, further data on optical properties (L*a*b* numerical coordinates and the reflectance for each wavelength in the visible spectrum) can be collected by VIS-RS. Figure 3 shows a comparison between reflectance curves acquired on tesserae belonging to Red, Yellow, Green, and Blue chromatic macrocategories identified by taking NCS-notations into account. Reflectance curves of tesserae belonging to the NCS-Red macrocategory displayed a very flat behaviour in the wavelength range between 400 and 580 nm, followed by an increase in reflectance intensity for the wavelengths above 580 nm. NCS-Blue and NCS-Green tesserae showed bell-shaped reflectance curves; the reflectance peak was located in the region between 440 and 540 nm for the blue tesserae, while, for the green tesserae, it was slightly shifted between 470 and 540 nm. Last, for the NCS-Yellow tesserae, reflectance curves were characterised by an increase in reflectance intensity for the wavelengths above 560 nm. Though further research is needed, the potentiality of VIS–RS seems to go far beyond the description of colours by means of reflectance curves and L*a*b* numerical coordinates. In particular, the shapes of the reflectance curves in the visible spectrum and the percentages of reflectance can deliver preliminary qualitative information relating to the colouring and opacifying agents [6,20,21].

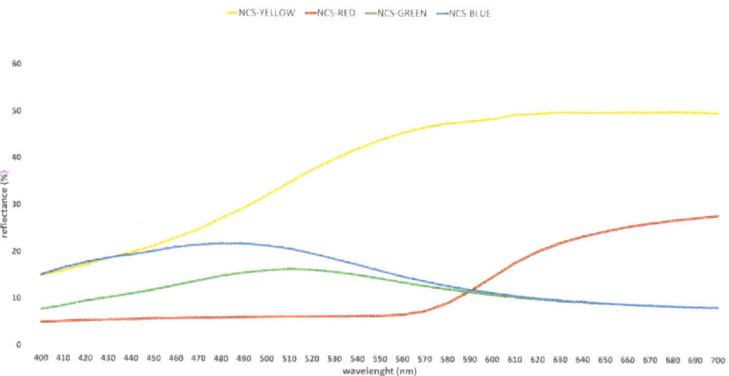

Figure 3. Comparison between reflectance curves acquired by VIS–RS on NCS-Red, Yellow, Green, and Blue tesserae.

After preliminary OM documentation, sampling, embedding, and polishing, SEM–EDS should be performed on deeply coloured glass. Back-scattered electrons (BSE) signals allow detecting and documenting of the different morphologies of the crystals precipitated into the glassy matrix, with EDS spot measurements also ascertaining their elemental composition. SEM–EDS is undoubtedly suitable to carry out high magnification morphological inspection of the inclusions dispersed in glassy matrix, as well as providing a qualitative and semiquantitative analysis of their elemental composition. However, to provide a more in-depth characterisation of these inclusions, necessary to identify raw materials responsible for the colour and opacity of the tesserae, SEM–EDS inspection needs to be integrated with other analytical techniques.

Especially if coloured and opaque, archaeological glasses are strongly heterogeneous materials; it is, thus, quite challenging to define what is the most suitable analytical technique for providing a full characterisation of the inclusions. It would, maybe, be more correct to claim that there is not only one. The choice of the technique/s is highly dependent upon the nature of the inclusions that we want to investigate. An integration of at least one molecular (such as Raman Microscopy) and one mineralogical analysis (X-ray Powder Diffraction—XRPD, micro-XRD or SEM equipped with an Electron Backscattered Diffraction detector—EBSD, if available) is recommended, offering an appropriate compromise to achieve a thorough characterisation of the inclusions responsible for the colour and opacity of glass.

The characterisation of lead–tin–antimony-based compounds found in ancient glasses as colouring and opacifying agent can be taken as an example. As discussed elsewhere in more detail [8], lead–tin–antimonate crystals have been attested to in several assemblages of opaque coloured glasses [22–25]. The most credited hypothesis is that the presence of lead–antimonate inclusions doped with tin could be related to the use of tin-rich metallurgical scraps [26,27]; therefore, an exact characterisation of these inclusions could represent the starting point for providing insights into the identification of a possible area of origin of the raw materials used as colouring and opacifying phases. SEM–BSE images (Figure 4a) provide information on the morphology of the inclusions, showing the occurrence of micrometric anhedral crystals in the vitreous matrix, frequently clustered together; EDS spot measurements carried out on the crystals gave preliminary information on the elemental composition of the crystals, demonstrating that they were mainly made of antimony and lead, although tin could also be detected (Figure 4b). To achieve an exact characterisation of the compound, Raman microscopy was directly performed on the inclusions (Figure 4c).

Acquired spectra showed the typical features assigned to lead antimonate doped with tin: in addition to the shifted Pb–O lattice mode at 140 cm^{-1}, a peak at about 450 cm^{-1}, an increase in the band at about 330 cm^{-1}, and a collapsed band at 510 cm^{-1} were observed, indicative of the partial replacement of the Sb^{+5} species by a larger Sn^{4+} cation [28,29].

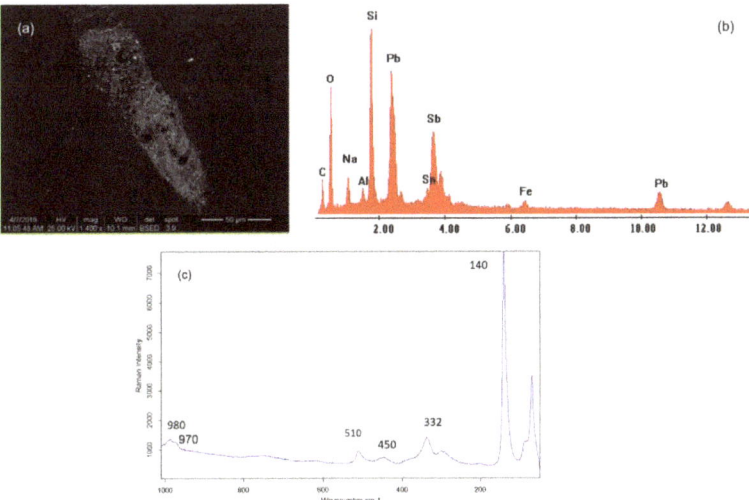

Figure 4. Example of lead–tin–antimony-based inclusions detected in a yellow mosaic glass tessera: (**a**) SEM–BSE image; (**b**) EDS spectrum; (**c**) Raman spectrum with bands at 140, 332, 450, and 510 cm^{-1} suggesting lead antimonate doped with tin [28,29].

In the above example, Raman microscopy can be considered as the most suitable analytical technique to gain data on the composition of the lead–tin–antimony-based compounds used to impart yellow colour and opacity to the glass. There are, however, cases in which the use of this analytical technique may not be informative. An example is represented by the opaque red coloured glass containing metallic copper inclusions. SEM–BSE images (Figure 5a) showed, at high magnification, nanometric rounded particles exclusively made of copper as EDS spot analysis demonstrate (Figure 5b). The nanometric size of these inclusions did not allow Raman microscopy analysis, where the magnifications of the most commonly accessible instruments hardly exceeds 1000×. Diffractometric analysis (Figure 5c) can, in this case, be decisive, allowing one to distinguish the presence of metallic copper within the glass matrix. It should, of course, be emphasized that, unless instruments such as microdiffractometers and/or BSDE detectors interfaced with SEM are available, diffractometric analysis usually requires powdering the sample and, therefore, is destructive.

Once the investigation of colouring and opacifying phases was completed, we analyzed the bulk chemistry. To investigate the base glass, both in terms of compositional recipes and provenance of raw materials, a combination of Electron Probe Micro Analysis (EPMA) and Laser Ablation–Inductively Coupled Plasma–Mass Spectrometry (LA–ICP–MS) analyses was proposed, together with specific data processing. Major and minor oxides, aimed at identifying both the fluxing agent and the "recipes" used in the glass-making process, can be determined by EPMA, while LA–ICP–MS analysis needs to be carried out for measuring trace elements and, thus, drawing inferences on the provenance of the sands used as vitrifying agents. Several analytical techniques can be employed for the quantification of major, minor, and trace elements in archaeological and historical glasses, such as Wavelength Dispersive X-ray Fluorescence spectrometry (WDXRF), Inductively Coupled Plasma–Optical Emission Spectrometry (ICP–OES), Ion Beam Analysis (IBA), and Neutron Activation Analysis (NAA) [30]. However, when dealing with opaque coloured glasses the prime advantage of combining EPMA and LA–ICP–MS for quantification of major to trace elements is the possibility of performing both analyses on the same mounted and polished samples where the study of colourants and opacifiers had previously been carried out. When dealing with archaeological and historical glasses, it is important to identify "recipes" that can be linked to primary production furnaces and, therefore, provide information on geographical areas of production and provenance [3]. However, for deeply coloured glass, a further degree of difficulty is due to the "contamination" of the compositional data of the matrix with materials added as colouring and opacifying agents.

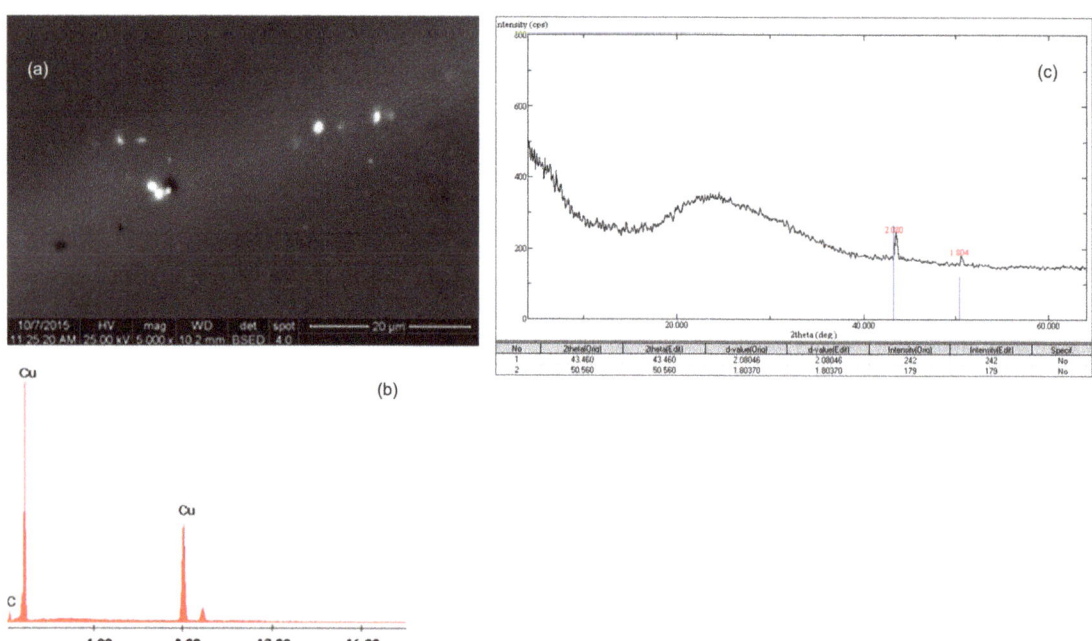

Figure 5. Example of nanometric copper-based inclusions detected in a red mosaic glass tessera: (**a**) SEM–BSE image; (**b**) EDS spectrum; (**c**) X-ray diffraction pattern showing the presence of metallic copper (Card No. standardreference4-0836).

To investigate the base glass, an integration between EPMA and LA–ICP–MS can work as a suitable compromise. Especially in the last few years, EPMA has been gradually being replaced by LA–ICP–MS for the determination of minor and major oxides, calculated by difference given a known oxide (generally SiO_2). Recent research shows, indeed, that close correspondence is generally observed between the data achieved by EPMA and "new generation" LA–ICP–MS equipment when analysing colourless or naturally coloured glasses [31,32]. LA–ICP–MS can perform major to trace element analysis of almost all elements within a sample during a single run, due to specific quantification protocols such as Internal Standard Independent (ISI) and Sum Normalization (SN) methods [33]. Although the potentialities of this technique are significant, its application to the study of deeply coloured opaque glass still needs to be thoroughly explored. In particular, it should be noticed that the most commonly used quantification method (the Sum Normalization) assumes that glass is almost exclusively comprised of oxides in known oxidation states and that the sum of the concentration of all oxides should equal 100% [33]. The first statement is quite difficult to verify when dealing with deeply coloured and opaque glasses such as tesserae, when the addition of several compounds is responsible for the colour shades and the opacity. Recent research also states that EPMA analysis of major and minor elements has several important advantages compared with other techniques, such as LA–ICP–MS as well as Secondary Ion Mass Spectrometry (SIMS): higher spatial resolution (from one to several μm), a well-established matrix correction procedure, and a lower cost [34]. Given the current state of knowledge, further studies should, therefore, be conducted on the specific application of LA–ICP–MS for the quantification of not only trace, but also major and minor elements in opaque deeply coloured glass before a complete dismission of EPMA for quantitative analysis of major and minor elements can be proposed. When dealing with opaque deeply coloured glasses, EPMA data can, however, be affected by the addition of materials acting as colourants and opacifiers. To better compare the composition of the base glass with the categories reported in the literature for naturally coloured glass, EPMA data should, thus, be adequately processed in order to, if not completely neutralise, at least minimize this effect of contamination. Following the method proposed by Robert Brill [35], reduced composition can be obtained by subtracting the oxides of elements presumably due to additives from the total sum of all those measured and by normalising the remaining data. However, the main concern associated with Brill's method remains the following: when the recalculation is carried out, how can we be sure not to incur arbitrary subtractions? Previously performed in-depth characterisation of colouring and opacifying phases can provide considerable help and a treasurable guide in preventing any subjective subtractions, as we do know what materials are responsible for the hue and opacity of the samples under study and this information can guide us in the recalculation of data.

4. Conclusions

The meticulous connection and correlation between archaeological, typological, and compositional data is a key requisite when approaching the study of archaeological glass. An actual coparticipatory approach between archaeologists and archaeometrists stands, therefore, as a fundamental requisite for understanding objects beyond their chemistry and recognising them as witnesses of human material culture.

In the attempt to outline a best practice approach for the study of archaeological glass, the reasons for comprehending the contexts as the first step toward a tailored selection of materials have been discussed. Understanding if we are in a context with sound evidence of production/working activities will lead to selection criteria other than those that will guide the selection of objects from multilayered contexts (or where no univocal evidence of production has been ascertained).

A proper discernment of the archaeological context will, therefore, support a well-fitted selection of the type(s) of finds to be subjected to chronotypological study and archaeometric investigation. The analysis of compositional and technological features cannot, in fact, be separated from a careful and detailed chronotypological study, aimed at framing the object and its morphology in the related chronological and geographical context of diffusion.

Regarding archaeometric analyses, it can be stated that the setup of the analytical protocol may vary depending on the features of the objects to be analysed. When dealing with naturally coloured glass, the analytical approach can be based on the use of fewer analytical techniques, the primary objective being the determination of the chemical composition of the base glass and, if needed, the provenance of the materials used as vitrifying agents. Otherwise, deeply coloured (and often opaque) glass shows heterogeneity in terms of microstructure and compositional features, due to the addition of colouring and opacifying agents. The setup of a more articulated analytical approach is, thus, needed, based on the interrelation of different investigation techniques to achieve an in-depth characterisation of colouring and opacifying agents.

When dealing with opaque coloured glasses, the NCS System proved to be a suitable tool for a preliminary documentation of colours among assemblages to be studied: since NCS-coordinates describe the hue of opaque coloured glass, they can provide a first split of objects/fragment to be analysed into what can be defined as chromatic macrocategories (i.e., green, blue, red, and black), avoiding any subjective classification and denomination of the colours and, consequently, facilitating comparison among objects from different assemblages. By carrying out VIS–RS measurements, more in-depth information on optical properties related to different chromatic shades among coloured glasses belonging to the same chromatic macrocategory can, then, be achieved.

Since opaque coloured glasses show extremely heterogeneous microstructures, their microstructural and microtextural features should be investigated before the composition of the glassy matrix, to reduce the risk of interferences and misunderstandings in data evaluation. SEM–EDS is the starting point for the characterization of microstructural and microtextural features: BSE observations enable one to investigate and document the morphologies of the crystals precipitated into the glassy matrix, and EDS spot measurements provide preliminary data on their elemental composition. SEM–EDS is, thus, undoubtedly suitable to carry out high-resolution morphological inspection of the inclusions dispersed in a glassy matrix, as well as a qualitative and semiquantitative analysis of their elemental composition. However, if a more in-depth characterisation of these inclusions is needed to identify raw materials responsible for the colour and opacity of the samples under study, SEM–EDS inspection needs to be integrated with other techniques. As discussed in this paper, the choice of the most suitable technique/s can vary, depending upon the nature of the inclusions to be investigated. It has, for instance, been demonstrated that μRaman can be extremely suitable for investigating the composition of lead–tin–antimony-based phases, but the same cannot be said for Cu-based nanoparticles, where mineralogical analyses can be more informative.

In-depth examination of the microstructure needs to be followed by a determination of the bulk chemistry of the tesserae under study. Major and minor oxides, whose measurement is aimed at identifying both the fluxing agent and the "recipes" used in the glass-making process, have been analysed by EPMA. LA–ICP–MS analysis was carried out to determinine trace elements to draw inferences on the provenance of the sands used as vitrifying agents.

To identify, with as much precision as possible, the compositional categories of the base glass the tesserae belong to, EPMA data should always be recalculated (and normalised) to minimise any effect caused by any other intentionally added compound. A characterisation of the colouring and opacifying agents carried out prior to the analysis of the base glass will, therefore, also allow avoiding (or at least minimising) any subjective subtraction.

Last but not least, it should be remembered that, when dealing with archaeological materials, the study approach can be susceptible to implementations and variations evaluated on two fundamental criteria: the material features of the objects under study, which will never be identical to one another, as they are the product of ancient technologies and not of industrial processes; technological advancement and all that the progress of research will offer us which could be suitable to further deepen our knowledge of this material category.

Author Contributions: Conceptualisation, S.F. and T.C.; data curation, S.F.; methodology, S.F. and T.C.; resources, M.V.; supervision, M.V.; validation, M.V.; writing–original draft, S.F. and T.C. All authors have read and agreed to the published version of the manuscript.

Funding: This research received no external funding.

Institutional Review Board Statement: Not applicable.

Informed Consent Statement: Not applicable.

Data Availability Statement: The study does not report new data.

Acknowledgments: The authors sincerely thank the anonymous reviewers for their comments and valuable suggestions, which have contributed to increase the quality of the manuscript.

Conflicts of Interest: The authors declare no conflict of interest.

References

1. Artioli, G. *Scientific Methods and Cultural Heritage: An Introduction to the Application of Materials Science to Archaeometry and Conservation Science*; Oxford University Press: Oxford, UK, 2010.
2. Edwards, H.; Vandenabeele, P. *Analytical Archaeometry: Selected Topics*; Royal Society of Chemistry: London, UK, 2012.

3. Rehren, T.; Freestone, I.C. Ancient glass: From kaleidoscope to crystal ball. *J. Archaeol. Sci.* **2015**, *56*, 233–241. [CrossRef]
4. Vandini, M.; Chinni, T.; Fiorentino, S.; Galusková, D.; Kaňková, H. Glass production in the Middle Ages from Italy to Central Europe: The contribution of archaeometry to the history of technology. *Chem. Pap.* **2018**, *72*, 2159–2169. [CrossRef]
5. Chinni, T.; Fiorentino, S.; Silvestri, A.; Vandini, M. Gambassini on the road. Underpinning evidence for a medieval widespread glass production in the north Adriatic area. *J. Archaeol. Sci. Rep.* **2021**, *37*, 103039. [CrossRef]
6. Fiorentino, S.; Vandini, M.; Chinni, T.; Caccia, M.; Martini, M.; Galli, A. Colourants and opacifiers of mosaic glass tesserae from Khirbet al-Mafjar (Jericho, Palestine): Addressing technological issues by a multi-analytical approach and evaluating the potentialities of thermoluminescence and optically stimulated luminescence da. *Archaeol. Anthropol. Sci.* **2019**, *11*, 337–359. [CrossRef]
7. Fiorentino, S. Fragile Connections, Persistent Methodology. A Tailor-Made Archaeometric Protocol to Investigate Technological and Cultural Issues in the Supply of Glass Tesserae under the Umayyad Caliphate. Ph.D. Thesis, University of Bologna, Bologna, Italy, 2019.
8. Vandini, M.; Fiorentino, S. From crystals to color: A compendium of multi-analytical data on mineralogical phases in opaque colored glass mosaic tesserae. *Minerals* **2020**, *10*, 609. [CrossRef]
9. Johnston-Feller, R. *Color. Science in the Examination of Museum Objects: Nondestructive Procedures*; The Getty Conservation Institute: Los Angeles, CA, USA, 2001.
10. Jarosewich, E. Smithsonian microbeam standards. *J. Res. NIST* **2002**, *107*, 681–685. [CrossRef]
11. Pouchou, J.; Pichoir, F. Determination of mass absorption coefficients for soft X-rays by use of the electron microprobe. In Proceedings of the 23rd Annual Conference of the Microbeam Analysis Society, Milwaukee, WI, USA, 8–12 August 1988; Newbury, D.E., Ed.; San Francisco Press: San Francisco, CA, USA, 1988; pp. 319–324.
12. Longerich, H.P.; Jackson, S.E.; Gunther, D. Inter-laboratory note. Laser ablation inductively coupled plasma mass spectrometric transient signal data acquisition and analyte concentration calculation. *J. Anal. At. Spectrom.* **1996**, *11*, 899–904. [CrossRef]
13. Kamber, B.S.; Greig, A.; Collerson, K.D. A new estimate for the composition of weathered young upper continental crust from alluvial sediments, Queensland, Australia. *Geochim. Cosmochim. Acta* **2005**, *69*, 1041–1058. [CrossRef]
14. Chinni, T. Produzione e Circolazione dei Manufatti In Vetro in Romagna nel Medioevo (V-XV sec.). Ph.D. Thesis, University of Bologna, Bologna, Italy, 2017.
15. Maltoni, S.; Chinni, T.; Cirelli, E.; Vandini, M.; Silvestri, A.; Molin, G. Archaeological and archaeometric study of the glass finds from the ancient harbour of Classe (Ravenna- Italy): New evidence. *Herit. Sci.* **2015**, *3*, 1–19. [CrossRef]
16. Maltoni, S.; Gallo, F.; Silvestri, A.; Vandini, M.; Chinni, T.; Marcante, A.; Molin, G.; Cirelli, E. Consumption, working and trade of Late Antique glass from north Adriatic Italy: An archaeometric perspective. In *Things That Travelled. Mediterranean Glass in the First Millennium CE*; Rosenow, D., Phelps, M., Meek, A., Freestone, I., Eds.; UCL Press: London, UK, 2018; pp. 191–214.
17. Silvestri, A.; Molin, G.; Salviulo, G. The colourless glass of Iulia Felix. *J. Archaeol. Sci.* **2008**, *35*, 331–341. [CrossRef]
18. Silvestri, A. The coloured glass of Iulia Felix. *J. Archaeol. Sci.* **2008**, *35*, 1489–1501. [CrossRef]
19. Sedláčková, H.; Rohanová, D.; Lesák, B.; Šimončičová-Koóšová, P. Medieval glass from Bratislava (ca 1200–1450) in the context of contemporaneous glass production and trade contacts. *Památky Archeol.* CV **2014**, 215–264.
20. Galli, A.; Poldi, G.; Martini, M.; Sibilia, E.; Montanari, C.; Panzeri, L. Study of blue colour in ancient mosaic tesserae by means of thermoluminescence and reflectance measurements. *Appl. Phys. A Mater. Sci. Process.* **2006**, *83*, 675–679. [CrossRef]
21. Galli, A.; Poldi, G.; Martini, M.; Sibilia, E. Thermoluminescence and visible reflectance spectroscopy applied to the study of blue-green mosaic silica-glass tesserae. *Phys. Status Solidi Curr. Top. Solid State Phys.* **2007**, *4*, 950–953. [CrossRef]
22. Maltoni, S.; Silvestri, A. A mosaic of colors: Investigating production technologies of roman glass tesserae from Northeastern Italy. *Minerals* **2018**, *8*, 255. [CrossRef]
23. Verità, M.; Maggetti, M.; Saguì, L.; Santopadre, P. Colors of Roman Glass: An Investigation of the Yellow Sectilia in the Gorga Collection. *J. Glass Stud.* **2013**, *55*, 21–34.
24. Schibille, N.; Neri, E.; Ebanista, C.; Ammar, M.R.; Bisconti, F. Something old, something new: The late antique mosaics from the catacomb of San Gennaro (Naples). *J. Archaeol. Sci. Rep.* **2018**, *20*, 411–422. [CrossRef]
25. Rosi, F.; Manuali, V.; Miliani, C.; Brunetti, B.G.; Sgamellotti, A.; Grygar, T.; Hradil, D. Raman scattering features of lead pyroantimonate compounds. Part I: XRD and Raman characterization of $Pb_2Sb_2O_7$ doped with tin and zinc. *J. Raman Spectrosc.* **2009**, *40*. [CrossRef]
26. Molina, G.; Odin, G.P.; Pradell, T.; Shortland, A.J.; Tite, M.S. Production technology and replication of lead antimonate yellow glass from New Kingdom Egypt and the Roman Empire. *J. Archaeol. Sci.* **2014**, *41*, 171–184. [CrossRef]
27. Lahlil, S.; Cotte, M.; Biron, I.; Szlachetko, J.; Menguy, N.; Susini, J. Synthesizing lead antimonate in ancient and modern opaque glass. *J. Anal. At. Spectrom.* **2011**, *26*, 1040–1050. [CrossRef]
28. Rosi, F.; Manuali, V.; Grygar, T.; Bezdicka, P.; Brunetti, B.G.; Sgamellotti, A.; Burgio, L.; Seccaroni, C.; Miliani, C. Raman scattering features of lead pyroantimonate compounds: Implication for the non-invasive identification of yellow pigments on ancient ceramics. Part II. in situ characterisation of Renaissance plates by portable micro-Raman and XRF studies. *J. Raman Spectrosc.* **2011**, *42*. [CrossRef]
29. Paynter, S.; Kearns, T.; Cool, H.; Chenery, S. Roman coloured glass in the Western provinces: The glass cakes and tesserae from West Clacton in England. *J. Archaeol. Sci.* **2015**, *62*, 66–81. [CrossRef]
30. Janssens, K. *Modern Methods for Analysing Archaeological and Historical Glass*; John Wiley & Sons Ltd.: Hoboken, NJ, USA, 2013.

31. Gratuze, B. Glass characterization using Laser Ablation-Inductively Coupled Plasma-Mass Spectrometry methods. In *Recent Advances in Laser Ablation ICP-MS for Archaeology*; Dussubieux, L., Golitko, M., Gratuze, B., Eds.; Springer: Amsterdam, The Netherlands, 2016; pp. 179–196.
32. Ceglia, A.; Cosyns, P.; Schibille, N.; Meulebroeck, W. Unravelling provenance and recycling of late antique glass from Cyprus with trace elements. *Archaeol. Anthropol. Sci.* **2019**, *11*, 279–291. [CrossRef]
33. Cagno, S.; Hellemans, K.; Janssens, K. The Role of LA-ICP-MS in the Investigation of Archaeological Glass. In *Recent Advances in Laser Ablation ICP-MS for Archaeology*; Dussubieux, L., Golitko, M., Gratuze, B., Eds.; Springer: Berlin, Germany, 2016; pp. 163–178.
34. Batanova, V.; Sobolev, A.; Magnin, V. Trace element analysis by EPMA in geosciences: Detection limit, precision and accuracy. *Mater. Sci. Eng.* **2018**, *304*, 1–17. [CrossRef]
35. Brill, R.H. *Chemical Analyses of Early Glass*; Corning Museum of Glass: New York, NY, USA, 1999.

Article

Non-Destructive pXRF on Prehistoric Obsidian Artifacts from the Central Mediterranean

Robert H. Tykot

Department of Anthropology, University of South Florida, Tampa, FL 33620, USA; rtykot@usf.edu; Tel.: +1-813-974-7279

Citation: Tykot, R.H. Non-Destructive pXRF on Prehistoric Obsidian Artifacts from the Central Mediterranean. Appl. Sci. 2021, 11, 7459. https://doi.org/10.3390/app11167459

Academic Editor: Marco Martini

Received: 11 July 2021
Accepted: 11 August 2021
Published: 13 August 2021

Publisher's Note: MDPI stays neutral with regard to jurisdictional claims in published maps and institutional affiliations.

Copyright: © 2021 by the author. Licensee MDPI, Basel, Switzerland. This article is an open access article distributed under the terms and conditions of the Creative Commons Attribution (CC BY) license (https:// creativecommons.org/licenses/by/ 4.0/).

Featured Application: The use of non-destructive, portable XRF instruments has expanded tremendously the elemental analyses of archaeological materials. This has revolutionized the data now available for our understanding of obsidian trade, maritime capabilities, and socioeconomic systems in the prehistoric central Mediterranean.

Abstract: Volcanic obsidian was widely used in ancient times for stone tools, with its highly glassy nature making it sharper than other lithics for cutting purposes. In Europe and the Mediterranean, there are just several island sources, and a few in one inland region, all having been used since the beginning of the Neolithic period, ca. 6000 BCE. Maritime transport was necessary for access to the Italian and Greek island sources, with the distribution of artifacts over distances up to 1000 km. While elemental analyses were used for identifying specific sources starting in the 1960s, the development of non-destructive and especially portable X-ray fluorescence spectrometers has revolutionized the number of artifacts tested since 2010, allowing statistically significant numbers for potential comparisons based on variables including time period, open-water distance, visual and physical properties, and cultural contexts. One overall accomplishment is the documentation of long-distance travel routes in the Tyrrhenian and Adriatic Seas of the central Mediterranean, based on the distribution proportions and quantity of obsidian artifacts from different geological sources, apparently following a down-the-line prehistoric exchange system. The spread of Palmarola obsidian is much greater than previously thought, while in Malta, Pantelleria obsidian was specifically selected for burial accompaniments on Gozo.

Keywords: obsidian; sourcing; trade and exchange; pXRF; trace elements; Italy; central Mediterranean; Neolithic; prehistory

1. Introduction

Volcanic obsidian was widely used in ancient times for stone tools, with its highly glassy nature making it sharper than other lithics for cutting purposes. In Europe and the Mediterranean, there are just several island sources in Italy and Greece, and a few in the inland Carpathian region, all having been used by the beginning of the Neolithic period, ca. 6000 BCE. Maritime transport was necessary for access to the island sources, while other materials would have been moved in opposite directions. Obsidian was likely a small part of this exchange system. The location and geological studies for each of the sources goes back to the early 20th century, while the ability to chemically identify specific sources began in the 1960s. The development of non-destructive and especially portable X-ray fluorescence spectrometers has revolutionized the number of artifacts tested since 2010, providing new interpretations of obsidian selection and use, which changed over time.

2. History of Obsidian

The modern word "obsidian" comes from the name of a Roman explorer, Obsius, who saw extensive geological quantities in Ethiopia [1], although obsidian sources on Italian islands, including Lipari and Sardinia and on the Greek island of Melos, had been used for thousands of years and were already well known from their wide usage.

Obsidian is an igneous glassy rock, formed during volcanic eruptions during the past 20 million years, which fractures conchoidally. It is typically an extrusive rock formed along the edges of viscous lava flows, in a volcanic dome, or when it cools while airborne. Some sources are intrusive, formed at the edges of dikes or sills. Obsidian may be found in many parts of the world, but only in certain geological regions where the magma composition was rhyolitic (or in some cases basaltic). Overall, the production of insufficient sizes to make stone tools (at least a few centimeters), the presence of gas vesicles trapped during formation (e.g., pumice), or the breakdown over time of unstable glass with phenocryst or spherulite crystalline formations limit the number of sources used in antiquity to certain mountainous terrestrial areas and volcanically formed islands (Figure 1).

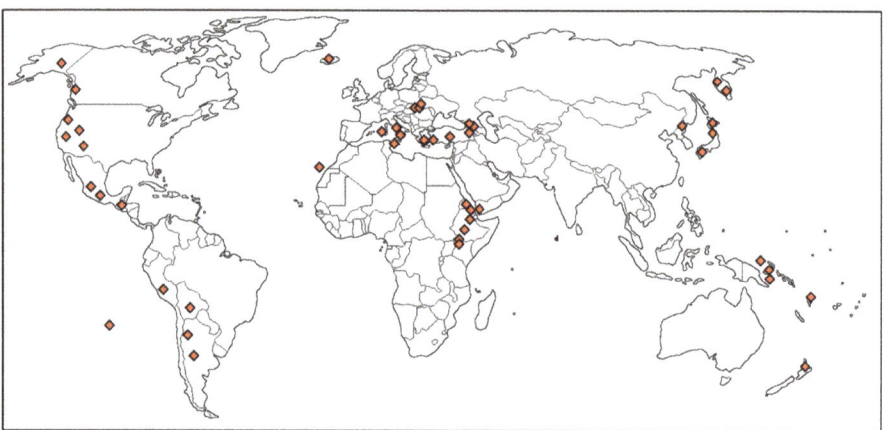

Figure 1. Map showing many of the obsidian source locations used in antiquity.

Obsidian was widely used for cutting and scraping tools during the Old and New Stone Ages, starting in the Lower Paleolithic by early *Homo* (ca. 1.75 million years ago) and continuing until recently in some parts of the world [2]. Due to its higher level of sharpness, compared to chert (flint), quartzite, and other stone tool material, it was intentionally acquired, flaked to produce broken edges, transported, and traded over distances of 1000 km or more within the central Mediterranean [3]. The preparation of cores was generally done near the geological source, with blades and other tools produced by trained lithic workers at many of the distant archaeological sites (Figure 2). Findings of obsidian artifacts at mainland sites in Greece and Italy, from island geological sources in the Mediterranean, infers the development of maritime travel by the late Upper Paleolithic [4] of at least simple crafts, to more complex vessels by the Early Bronze Age (3rd millennium BCE); the earliest remains found in the Mediterranean of open-water boats (or drawings) are only more recent. Obsidian is still used today for surgical procedures (including eye and heart), due to its greater sharpness and faster healing process, compared to stainless steel scalpels (just search online for a commercial scalpel vendor). Obsidian was also used for polished mirror surfaces, and for jewelry in certain cultures even today.

Figure 2. Example of Neolithic obsidian blade tools from Sicily. Scale in centimeters. Most stone tools would have been mounted on wood or bone handles (rarely preserved).

In addition to the size of natural obsidian blocks, and the quantity produced in a geological source area, the visual characteristics of obsidian were important, especially when there was a variety in color, luster, transparency, and the presence of phenocrysts. Most obsidian is in the black-to-gray color range, but there are some sources with brown, tan, red, orange, yellow, or blue, often mixed with black, caused by some inclusions in the magma or by trace elements. The orientation of any patterns may be indicative of the geological formation process (e.g., lava flow).

The ability to identify the geological origin of obsidian artifacts found at archaeological sites allows the reconstruction of cultural interaction and trade patterns, including the likely movement of other materials (e.g., pottery, domesticated animals, clothing, food products) as well.

3. Analytical Methods

The chemical analysis of obsidian to look at trace elements began in the early 1960s with the use of optical emission spectrometry, with simple X-Y graphs, such as barium vs. zirconium, distinguishing many sources in the Mediterranean and Near East [5]. By the mid-1970s, additional methods, including X-ray fluorescence (XRF) spectrometry and instrumental neutron activation analysis (INAA), were able to produce even more distinctive results by using many trace elements with excellent precision [6,7]. In addition, fission-track dating was also used, discriminating the limited numbers of obsidian sources by differences in their formation ages [8]. During the 1980s, major and minor elements were also shown to be successful in distinguishing obsidian sources, using atomic absorption spectroscopy (AAS) [9] and scanning electron microscopy with an energy-dispersive spectrometer (SEM-EDS) [10]. The homogeneity in the composition of molten lava, and the rapid formation of glassy obsidian, have led to significant differences between sources in composition for many major and trace elements. This allows a choice of simple X-Y graphs of a few elements to distinguish source groups.

Despite the development of these many methods of successful analyses by the 1980s, little detailed research had been done on the geological obsidian sources from an archaeological perspective, particularly that discriminating between multiple subsources of usable obsidian within each island. The total number of central Mediterranean obsidian artifacts that had been analyzed in the 1960s and 1970s was less than 200, and by the end of the 1980s, this had reached only about 500 total (see table IX, pp. 66–68 in [11]). Many of those artifacts tested had come from museum collections dating back to the late 19th and early 20th centuries, not having archaeological contexts or chronology. Even by 1995, the total analyzed was fewer than 900, with only 26 sites having 10 or more artifact analyses and allowing statistical comparisons (Table 1). Nevertheless, a lot had been learned about obsidian trade by that time [12].

Table 1. Analyses of obsidian artifacts in the central Mediterranean.

Years	$\geq$10 Per Site	$\geq$25 Per Site	Analyses	% Total
1964–1994	26	7	884	5%
1995–2010	97	33	3217	19%
2011–2021	158	108	12,895	76%
Total	281	148	16,996	100%

In the 1990s, however, the introduction of inductively coupled plasma (ICP), optical emission spectrometry (OES), and mass spectrometry (MS), often with laser ablation (LA), allowed the minimally destructive analysis of artifacts and increased the numbers tested [11,13–15]. XRF also developed further, with some instruments allowing minimally destructive (or even non-destructive) analyses. The use of mounted microsamples of 1–2 mm wide solid pieces of obsidian, with as many as 16 samples on a single 1-inch disk, was developed for electron probe microanalysis (EPMA) for this reason [11,13]. Similar analyses were conducted using a scanning electron microscope (SEM-EDS) on obsidian artifacts [16]. Overall, the use of automated analyses also reduced costs, which are always a limiting factor for archaeological studies. By 2010, more than 3000 obsidian artifacts in the central Mediterranean had been analyzed [13].

Starting in this millennium, the creation of desktop, non-destructive XRF analysis instruments [17–19], and especially portable XRF spectrometers, has revolutionized the analysis of obsidian in many parts of the world [20–26]. The homogeneity of obsidian and its high resistance to weathering are part of the success of non-destructive surface analyses. The pXRF may simply be placed adjacent to the cleaned object in the field, running on batteries and using a built-in computer, or within museums or other facilities (Figure 3). The main reasons for initial commercial production of pXRFs was for businesses and agencies with applications such as field testing of soils near factories; however, its utility for analyzing art and archaeology objects represented an academic market, too. After about 2005, commercially produced hand-held portable XRF (pXRF) instruments were available from several companies. Their small size makes them transportable in a backpack and as carry-on luggage on planes.

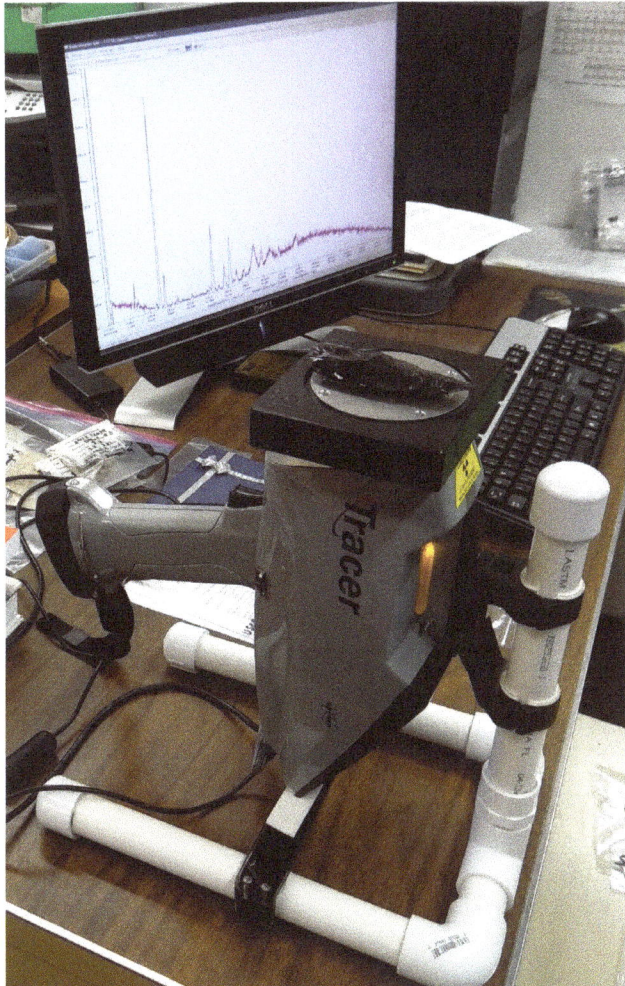

Figure 3. Most recent pXRF instrument (Bruker Tracer Vg) used in these studies. Mounted upright using a home-made plastic stand.

In traditional XRF instruments, samples are placed within a vacuum chamber so that secondary X-rays are not absorbed prior to reaching the detector. However, a vacuum is not necessary for quantitative measurements of elements above potassium, so that obsidian objects need not be contained in a vacuum chamber for measurement of K-shell electron replacements of elements from Ca to La. For hand-held XRF instruments, 50 kV is the highest energy setting for primary X-rays, thus limiting K-line energy measurements of only a few elements in periodic table row 6. For pXRF analyses, elements specifically used for obsidian source identification include major elements Ca, Ti, Mn, Fe, and trace elements, including Rb, Sr, Y, Zr, and Nb (Figure 4). The use of a filter (12 mil Al, 1 mil Ti, 6 mil Cu) reduces the background for these elements, with detection limits for trace elements in single digit ppm [27,28].

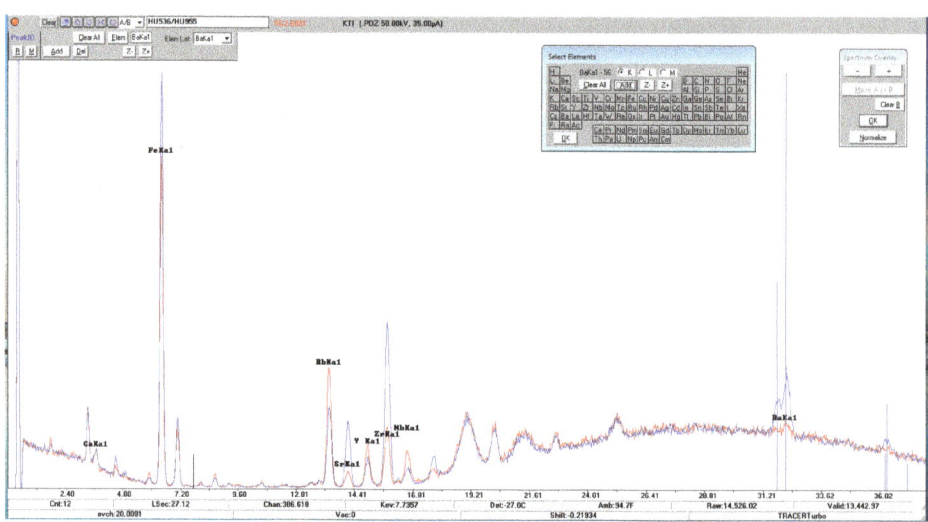

Figure 4. K-line energy peaks for two obsidian samples. In blue is from Monte Arci (Sardinia) SC; note the high strontium, zirconium, and barium peaks when compared to SA (in red). The blue lines represent the multiple peaks for barium.

As with all XRF instruments, the specific energy peaks for different elements, especially the L-lines for elements in period table rows 6 and 7 (e.g., Cs, Ba, La, Ce, Nd, Th, U) overlap with the K-lines for lower Z elements, thus limiting precision for those trace elements with similar energy levels to major elements (e.g., Fe and below) when significantly present. Other elements, such as those used for compositional analysis of metals and other materials (Ti, Cr, Fe, Co, Ni, Cu, Zn, As, Pd, Ag, Cd, Sn, Sb, Pt, Au, Hg, and Pb), may also be detected, and other filters used to minimize background effects [29].

In most cases, the highly precise measurements of the K-lines of just a few of these trace elements is sufficient to distinguish obsidian sources in any part of the world, while multi-variate statistics of 5–7 trace elements may be used for identifying pottery production groups [27]. The beam size of the X-rays is typically about 5–8 mm in diameter, with some having options for smaller beams reaching the sample. The length of time necessary for analysis has decreased with newer pXRF models, which use silicon drift (SDD) rather than silicon PIN (Si-PIN) detectors; with the first Bruker Tracer model used in 2007–2012 (III-V+), each spot for trace element analysis was run for 180 s; for 2013–2016 (III-SD), 60–90 s; and for 2017 to the present (Vi, Vg), just 30 s. Running for longer times does not go beyond limitations of the detector and increase the precision or sensitivity limits. For homogenous obsidian, one analysis per artifact was sufficient and only in a few cases were reruns needed to confirm the results and source assignment. In some of those cases, the samples were of irregular shape, or were small bladelets just a few mm in width and fewer in thickness, with lower than usual total counts. Incorrect source assignment is avoided in such cases by using trace element ratios.

Museums and government agencies in many countries are open to international collaboration and access to their archaeological collections, with analytical research facilitated when done without necessary permissions for destructive sampling, and movement (of intact artifacts or samples) to laboratories, even within the same country. The portability, easy operation, and low maintenance for hand-held XRFs also enabled many archaeologists to conduct analyses themselves, without depending on a geoscientist or laboratory staff to prepare and run their samples. The homogeneity of obsidian, the relatively flat areas on stone tools that may be tested, and little if any surface weathering make it a perfect material for non-destructive X-ray fluorescence analysis.

For a number of years, however, there were concerns raised about the integrity of analyses being conducted and, in particular, the production of data with actual concentration values [23,30]. How do we deal without calibrated quantitative results for major elements silicon (typically 65–75% for SiO_2 in obsidian), aluminum, sodium, potassium, and magnesium? How do we deal with matrix effects on secondary X-rays? What standard reference materials may be used for inter-laboratory comparisons? These issues were mostly for archaeology users who needed to compare their analytical data for artifacts with those of other scholars' analyses of geological source samples in their geographic region. Obsidian calibration software was developed and shared, however, by 2008 for the Bruker pXRF instruments, based on 40 geological obsidian samples analyzed by INAA, LA-ICP-MS, and XRF [22,31]. Separately, pXRF users who analyzed sufficient geological samples of known origin could also make a direct comparison of the uncalibrated raw data with that from the archaeological samples that they analyzed with the same instrument. Nevertheless, the use of calibrated data is expected for publications in many journals. For archaeological obsidian artifact studies, nearly all research is conducted by scholars with geological samples from their region of interest.

4. Obsidian Sourcing in Europe and the Mediterranean

There are four obsidian sources in the Central Mediterranean, from volcanic formations on the Italian islands of Sardinia [11,13,32–35], Palmarola [36], Lipari [37,38], and Pantelleria [3,39–42] (Figure 5). People with agriculture lived on both Sardinia and Lipari from the start of the Early Neolithic (ca. 6000 BCE), with the use of obsidian for stone tools starting at the same time. No one settled on the tiny island of Palmarola, while it appears that Pantelleria was not occupied until about 3000 BCE [43]. There is also Melos in the Aegean with two subsources [44], and one Carpathian source that was used, in southeastern Slovakia [45,46]. Geological surveys and analytical research were conducted on each, assessing the quantity and quality of obsidian from multiple outcrops and the ability to distinguish between these sources. For the four Italian islands, analyses were conducted by the author, using INAA, LA-ICP-MS, and EDS-XRF, defining multiple subsources for each island, which is important for the study of archaeological artifacts and our interpretation of prehistoric access and collection of obsidian [2,3,35,47].

Non-destructive analyses by pXRF of obsidian artifacts are also able to distinguish Monte Arci (Sardinia) subsources (Sardinia A, Sardinia B1, Sardinia B2, Sardinia C), as well as for Lipari (Gabellotto, Canneto Dentro, Monte Guardia) and Pantelleria (Lago di Venere 1 and 2, Balata dei Turchi) (Figures 6–9). At least with the elements currently measured and calibrated, we cannot distinguish the three Balata dei Turchi subgroups, nor the three subsources on Palmarola. Given the circumstances in which obsidian would have been obtained on these two islands during the Neolithic period, in particular by visitors rather than residents, these distinctions are not considered important.

Figure 5. Map showing obsidian sources in Europe and the Mediterranean.

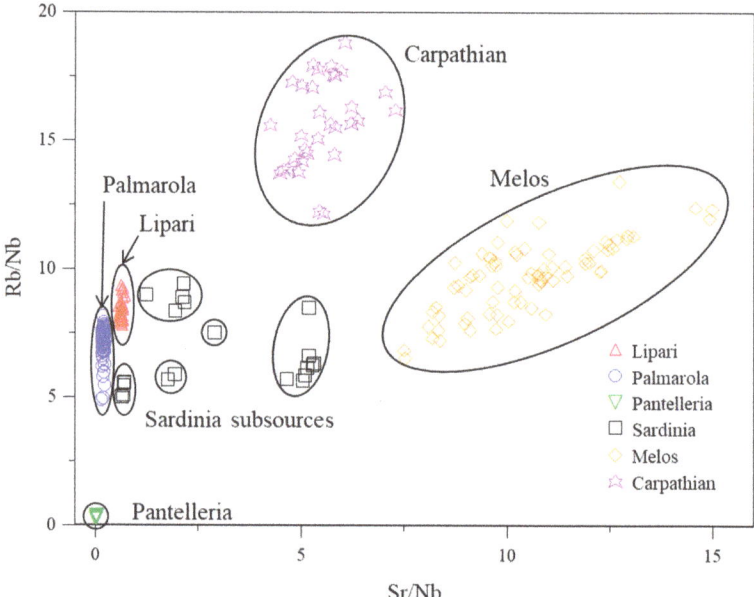

Figure 6. Distinguishing European obsidian sources with selected trace elements. Not all geological samples tested are shown.

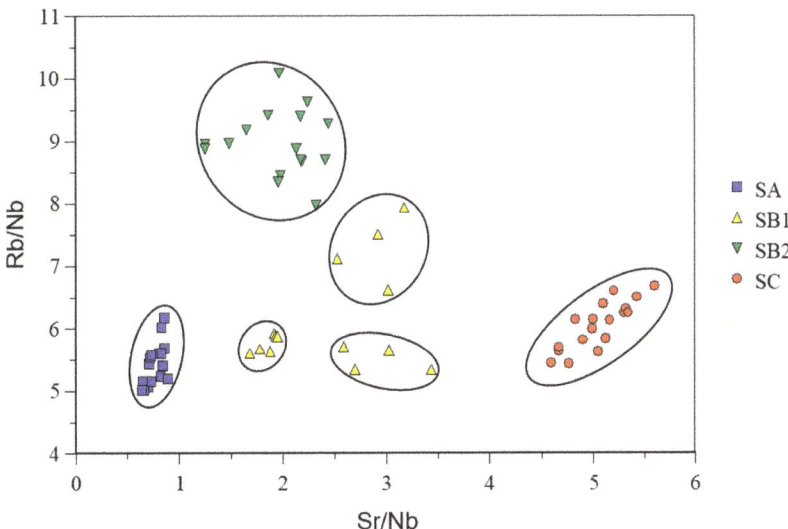

Figure 7. Separation of multiple subsources for Monte Arci (Sardinia) using a selection of geological samples collected by the author.

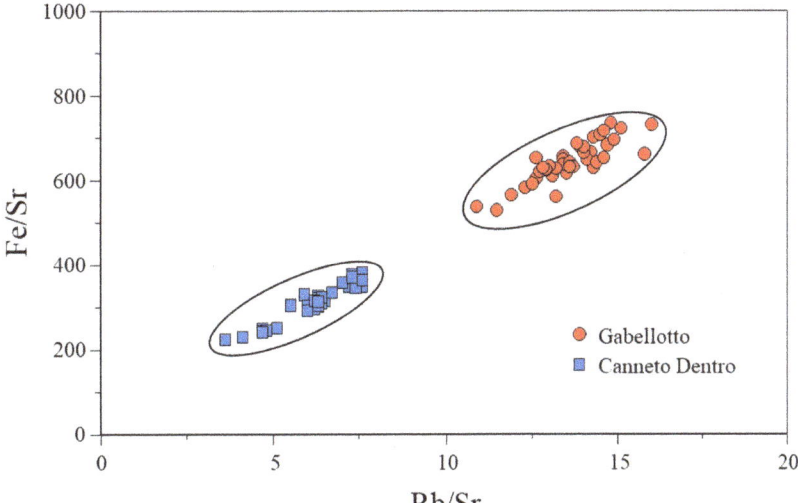

Figure 8. Two prehistoric Lipari obsidian groups used for tools, using geological samples collected by the author.

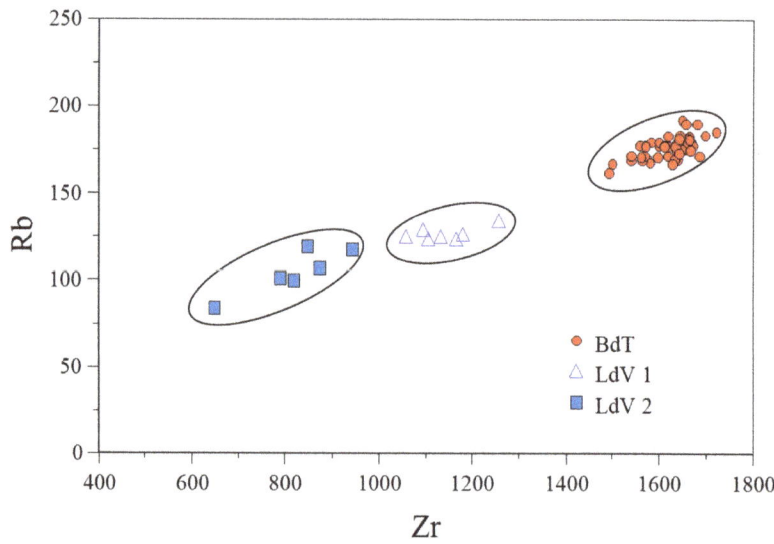

Figure 9. Pantelleria is separated by Balata dei Turchi (BdT), and both Lago di Venere (LdV) 1 and 2.

5. Applications of pXRF on Obsidian Artifacts from Central Mediterranean Sites

With a non-destructive pXRF, the author expanded his research on obsidian, from a focus on Sardinia and Corsica [13,33,42,48–50] to throughout Italy as well as to Malta and Croatia [3,51–54]. In the past ten years, more than 12,000 artifacts have been analyzed, with ≥ 25 from each of the > 100 archaeological sites (Figure 10).

5.1. Sardinian Obsidian in Continental Italy

The transportation and trade of obsidian from Monte Arci in Sardinia to Corsica and beyond was realized in the early study by Hallam et al. [7], which identified it at several sites in Southern France and Northern Italy. Since then, Monte Arci obsidian has been identified at many sites in Northern Italy, which also is a great distance from the other central Mediterranean sources, with changes over time in proportion to Lipari obsidian [55,56]. In addition, social network analysis supports hypotheses of different obsidian pathways, including open-water north-bound from Corsica to Southern France [57]. With the very large number of non-destructive analyses conducted in this millennium, we now see that Sardinian obsidian also reached central Italy as a significant percentage of assemblages and made its way to southernmost Italy and even Sicily in very small numbers, supporting an interpretation of a down-the-line type exchange during the Neolithic [3]. A total of just 23 artifacts of Sardinian obsidian (out of nearly 1300 obsidian artifacts analyzed) were identified at sites south of Rome, indicating that travel routes were not directly across the Tyrrhenian, but from Sardinia to Corsica, then through the Tuscan archipelago and southward, on or along the Italian peninsula (Figure 11). Undoubtedly, other materials were exchanged and traveled in opposite directions, including domesticated animals (sheep, goat, cattle, pig), produce (from wheat, barley, other plants), clothing, tools, wood, etc. [58].

Figure 10. Map of central Mediterranean showing archaeological sites with ≥10 source analyses of obsidian artifacts. Sites in red analyzed by the author.

Figure 11. Map with sites in southern Italy with Sardinia obsidian artifacts. From north to south, in red circles: Poggio Olivastro 20/100 (20%); Casale del Dolce 1/35 (3%); Venafro 1/132 (<1%); Pulo di Molfetta 1/37 (2.7%); M. Di Gioia 12/12 (100%); Ausino 3/21 (14%); Saracena 4/842 (<1%); Bova Marina 1/200 (<1%); Valdesi 1/41 (2.5%).

5.2. Obsidian Artifacts from Sites around the Adriatic and in Croatia

Prior to its political breakup, no obsidian artifacts found in Yugoslavia had been analyzed. Since then, however, a significant amount of archaeological research was conducted in Croatia, with obsidian found at many prehistoric sites along the Dalmatian coast, on islands in the Adriatic Sea, and as far north as Istria. At first, the results of analyses by pXRF, which showed that most were from Lipari, with very little reaching there over land from the Carpathian sources, were not expected [53,59]. In addition, the presence of some from Palmarola, on the island of Sušac and on the mainland at Lok. Musa, was surprising considering our previous thoughts of it having limited distribution [47] (Figure 12). Obsidian from Palmarola, however, was identified at three Neolithic sites along the Italian coast as well as at sites in the Tavoliere and near the Gargano Peninsula from where travel by island hopping over the Adriatic likely occurred [47,53,60]. Entirely unexpected was the identification of four obsidian artifacts on the island of Palagruža coming from Melos

(Sta Nychia subsource) since no others have been securely identified in prehistoric contexts west of Albania [59,61]. The Palagruža site, however, dates to the Copper Age, when increasing social and economic complexity began in the Aegean.

Figure 12. Map with directions to archaeological sites with Palmarola obsidian artifact attributions, including across the Adriatic.

5.3. The Small Island of Ustica, Northwest of Sicily

Ustica is a small island, about 8 km^2, located > 50 km north of Capo Gallo (coast north of Palermo in western Sicily). Despite its isolated location, there is archaeological evidence of its occupation also beginning in the Neolithic time period, indicating the rather long-distance open-water travel capabilities at that time. The use of obsidian artifacts, coming from both Pantelleria and Lipari, was first shown with analyses of artifacts from excavations of a Bronze Age site [62], and more recently at several different sites with more than 1100 obsidian artifacts found and analyzed [51,63–65] (Figure 13). In comparison to travel of much more modest distances from northern Corsica through the Tuscan archipelago to mainland Italy, and across the central Adriatic to Croatia, Ustica was not a stop along the way to other places but instead the only reason for travel in its direction.

As noted already, there is no direct or indirect evidence of the boats, rafts, or other open-water vessels that were used in the 6th millennium BCE to reach islands in the Mediterranean (or in the following 4000 years), but it strongly appears that such travel was regular and able to transport not only a few people but also domesticated animals and other items of their material culture. Obsidian from Lipari, in the Aeolian Islands, may have been transported over 25 km to northeastern Sicily, then along the shorelines to Palermo and from there to Ustica, but the proportions of Lipari to Pantelleria obsidian (average of 90% Lipari, 10% Pantelleria), when compared to what was found at the site of Grotta dell'Uzzo in northwest Sicily, suggest more direct open-water travel of >100 km from the westernmost Aeolian island of Alicudi to Ustica [65]. This is supported by the open-water distance from Pantelleria to southwest Sicily (~100 km), followed by travel along the western coast and then > 50 from Capo Gallo to Ustica. The small but seemingly consistent percentage of obsidian from Pantelleria reflects not only the much greater distance of travel, but also its visual and physical properties and their demand when compared with Lipari obsidian.

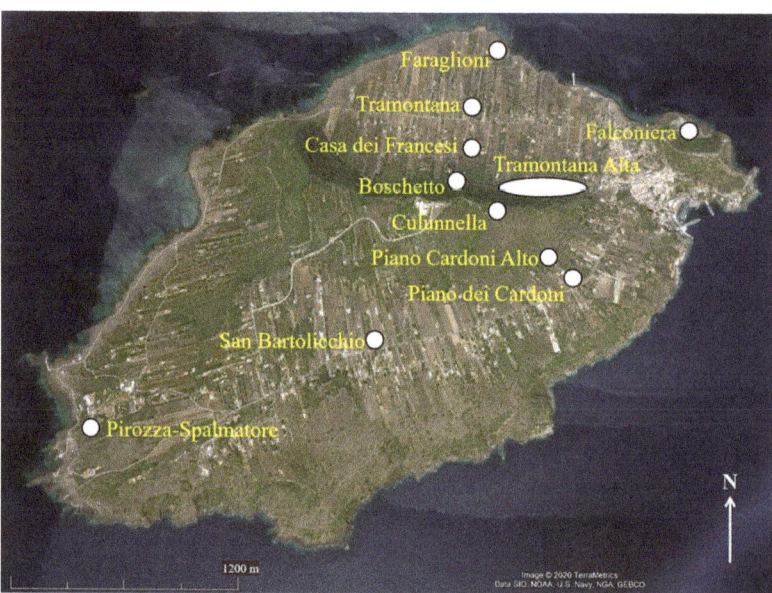

Figure 13. Archaeological sites on Ustica (north of Palermo, Sicily). The ellipse for Tramontana Alta is the sloped pathway along which surface artifacts were collected.

5.4. Obsidian at Sites in Malta

The Maltese Islands are about the same distance south of Sicily as Pantelleria, while they were occupied from the Neolithic, and, like elsewhere, were using obsidian from both Lipari and Pantelleria [3,66]. Three different sites were excavated: Skorba, occupied throughout the Neolithic, and Tas-Silj, which is Bronze Age, both on Malta; and the Brochtorff Circle at Xaghra, on the smaller island of Gozo, which is Copper Age (Figure 14).

Skorba has a combination of residential and ritual structures, with seven phases spanning from ca. 5500–2500 BCE (Ghar Dalam, Grey Skorba, Red Skorba, Zebbug, Ggantija, Saflieni, and Tarxien) and each with obsidian present [67]. Following the excavation, visual distinctions of the nearly 300 obsidian artifacts found were initially used to assign nearly 80% to Lipari (black-grey) and 20% to Pantelleria (dark green), and a selection of 25 were analyzed by INAA to confirm this [7]. More recent analyses by pXRF confirmed the visual assignments while also indicating that those from Pantelleria mostly came from Balata dei Turchi with a small number from Lago di Venere, and that all of the Lipari obsidian came specifically from Gabellotto. The proportion of Lipari to Pantelleria for each time period was consistent, and quite similar to that for the sites on Ustica.

Much more recent excavations at the Brochtoff Circle at Xaghra revealed a large number of underground, individual chamber tombs dating to the 3rd millennium BCE [68]. More than 100 obsidian artifacts were analyzed by pXRF, with the results being very different than at the site of Skorba, overall. Only 28% were from Lipari and 72% from Pantelleria, while at Skorba, none of the small number (8) of obsidian artifacts from the contemporary Tarxien phase came from Pantelleria. Again, all of the Lipari obsidian was from the Gabellotto subsource, while the Pantelleria artifacts came from multiple subsources on Pantelleria, mostly Balata dei Turchi. Since the many tomb chambers, representing about 700 individuals, were not of the same time period but spanned many families over at least a few hundred years, the high selection of Pantelleria obsidian was not a single incident, but an extended cultural burial practice.

Figure 14. Malta island sites with obsidian artifacts.

6. Discussion and Conclusions

These four examples of obsidian studies in the central Mediterranean illustrate the importance of analyzing large numbers of artifacts, allowing statistically significant numbers for comparisons based on variables, including time period, open-water distance, visual and physical properties, and cultural contexts. One overall accomplishment is the documentation of long-distance travel routes, based on the distribution proportions and quantity of obsidian artifacts from the different geological sources (Figure 15). This involved minimizing open-water travel when possible, while demonstrating multiple steps and the likely multiple transactions during the Neolithic, prior to the development of both complex societies and technological advancements in the Bronze Age, leading also to the decreased long-distance distribution of obsidian. In general, there was a major long-distance distribution to the north and northwest, which began with the spread of agriculture and domesticated animals in the Early Neolithic and continued through the end of the Late Neolithic (ca. 6000–3000 BCE) [2,3,37].

Over the last decade, more than 75% of the elemental analyses of obsidian were accomplished with the development of non-destructive XRF instruments, especially those that are hand-held and easily portable, which have enabled low-cost and rapid analyses of archaeological artifacts within museums and storage facilities. This has changed the status of our research on obsidian, as described in the recent past [14]. Many other methods have shown to be quite successful in distinguishing geological sources, and there are still experimental studies continuing with a variety of analytical methods, including Cl and Na proportions, geological formation age, and magnetic properties. Most important, however, is still the need for further studies of excavated obsidian artifacts, for different time periods and geographic locations, and integration with studies of lithic typology, the technology used for the production of stone tools, microscope-based use-wear patterns, and other parts of the *chaîne opératoire*.

Figure 15. Obsidian distribution directions in the central Mediterranean.

Funding: Travel costs for this research have been partly funded by the University of South Florida.

Institutional Review Board Statement: Not applicable.

Informed Consent Statement: Not applicable.

Data Availability Statement: Elemental data for individual artifacts is published or in course of publication. Requests may be sent to the author.

Acknowledgments: The extensive use of the pXRF would not have been possible without the consent of many colleagues and permissions obtained from many museums and government superintendencies in Sicily, peninsular Italy, Croatia, and Malta. I thank Andrea Vianello for assistance with this, as well as with the use of the pXRF in Italy. I also thank Kyle Freund for assistance in conducting analyses in Sicily and Southern Italy. The pXRF instruments were acquired by the author.

Conflicts of Interest: The author declares no conflict of interest.

References

1. Pliny the Elder (77 AD). *Naturalis Historia Book 36*. Chapter 67. Available online: http://www.perseus.tufts.edu/hopper/text?doc=Plin.+Nat.+toc (accessed on 1 August 2021).
2. Tykot, R.H. Obsidian in prehistory. In *Encyclopedia of Glass Science, Technology, History, and Culture*; Richet, P., Ed.; John Wiley & Sons, Inc.: Hoboken, NJ, USA, 2021.
3. Tykot, R.H. Obsidian studies in the prehistoric central Mediterranean: After 50 years, what have we learned and what still needs to be done? *Open Archaeol.* **2017**, *3*, 264–278. [CrossRef]
4. Jacobsen, T.W. Excavation in the Franchthi Cave, 1969–1971, Part I. *Hesperia J. Am. Sch. Class. Stud. Athens* **1973**, *42*, 45–88. [CrossRef]

5. Cann, J.R.; Renfrew, C. The characterization of obsidian and its application to the Mediterranean region. *Proc. Prehist. Soc.* **1964**, *30*, 111–133. [CrossRef]
6. Belluomini, G.; Taddeucci, A. Studi sulle ossidiane italiane. III—Elementi minori. *Period. Miner.* **1971**, *40*, 11–39.
7. Hallam, B.R.; Warren, S.E.; Renfrew, C. Obsidian in the western Mediterranean: Characterisation by neutron activation analysis and optical emission spectroscopy. *Proc. Prehist. Soc.* **1976**, *42*, 85–110. [CrossRef]
8. Bigazzi, G.; Bonadonna, F.P.; Belluomini, G.; Malpieri, L. Studi sulle ossidiane italiane. IV. Datazione con il metodo delle tracce di fissione. *Boll. Soc. Geol. Italy* **1971**, *90*, 469–480.
9. Michels, J.; Atzeni, E.; Tsong, I.S.T.; Smith, G.A. Obsidian hydration dating in Sardinia. In *Studies in Sardinian Archaeology*; Balmuth, M.S., Rowland, R.J., Jr., Eds.; University of Michigan: Ann Arbor, MI, USA, 1984; pp. 83–113.
10. Biró, T.K.; Pozsgai, I.; Vlader, A. Electron beam microanalyses of obsidian samples from geological and archaeological sites. *Acta Archaeol. Acad. Sci. Hung.* **1986**, *38*, 257–278.
11. Tykot, R.H. Prehistoric Trade in the Western Mediterranean: The Sources and Distribution of Sardinian Obsidian. Ph.D. Thesis, Department of Anthropology, Harvard University, Harvard, MA, USA, 1995.
12. Williams-Thorpe, O. Obsidian in the Mediterranean and the Near East: A provenancing success story. *Archaeometry* **1995**, *37*, 217–248. [CrossRef]
13. Tykot, R.H. Characterization of the Monte Arci (Sardinia) Obsidian Sources. *J. Archaeol. Sci.* **1997**, *24*, 467–479. [CrossRef]
14. Tykot, R.H. Scientific methods and applications to archaeological provenance studies. In *Physics Methods in Archaeometry. Proceedings of the International School of Physics "Enrico Fermi" Course CLIV*; Martini, M., Milazzo, M., Piacentini, M., Eds.; Società Italiana di Fisica: Bologna, Italy, 2004; pp. 407–432.
15. Tykot, R.H.; Glascock, M.D.; Speakman, R.J.; Atzeni, E. Obsidian subsources utilized at sites in southern Sardinia (Italy). In *MRS Online Proceedings Library (OPL), Volume 1047: Symposium Y—Materials Issues in Art and Archaeology VIII*; Vandiver, P.B., McCarthy, B., Tykot, R.H., Ruvalcaba-Sil, J.L., Casadio, F., Eds.; Cambridge University Press: Cambridge, UK, 2008; pp. 175–183.
16. Le Bourdonnec, F.-X.; Bontempi, J.-M.; Marini, N.; Mazet, S.; Neuville, P.F.; Poupeau, G.; Sicurani, J. SEM-EDS characterization of western Mediterranean obsidians and the Neolithic site of A Fuata (Corsica). *J. Archaeol. Sci.* **2010**, *37*, 92–106. [CrossRef]
17. De Francesco, A.M.; Crisci, G.M.; Bocci, M. Non-destructive analytical method by XRF for the determination of the provenance of archaeological obsidians from the Mediterranean area. A comparison with *traditional* XRF method. *Archaeometry* **2008**, *50*, 337–350. [CrossRef]
18. De Francesco, A.; Bocci, M.; Crisci, G.M. Non-destructive applications of wavelength XRF in obsidian studies in X-ray fluorescence spectrometry (XRF) in geoarchaeology. In *X-ray Fluorescence Spectrometry (XRF) in Geoarchaeology*; Shackley, M.S., Ed.; Springer: New York, NY, USA, 2011; pp. 81–107.
19. De Francesco, A.M.; Bocci, M.; Crisci, G.M.; Francaviglia, V. Obsidian provenance at several Italian and Corsican archaeological sites using the non-destructive X-ray fluorescence method. In *Obsidian and Ancient Manufactured Glasses*; Liritzis, I., Stevenson, C.M., Eds.; University of New Mexico Press: Albuquerque, NM, USA, 2012; pp. 115–130.
20. Cecil, L.G.; Moriarty, M.D.; Speakman, R.J.; Glascock, M.D. Feasibility of field-portable XRF to identify obsidian sources in central Peten, Guatemala. In *Archaeological Chemistry. Analytical Techniques and Archaeological Interpretation*; Glascock, M.D., Speakman, R.J., Popelka-Filcoff, R.S., Eds.; American Chemical Society: Washington, DC, USA, 2007; pp. 506–521.
21. Craig, N.; Speakman, R.J.; Popelka-Filcoff, R.S.; Glascock, M.; Robertson, J.D.; Shackley, M.S.; Aldenderfer, M.S. Comparison of XRF and PXRF for analysis of archaeological obsidian from southern Peru. *J. Archaeol. Sci.* **2007**, *34*, 2012–2024. [CrossRef]
22. Ferguson, J.R. X-ray fluorescence of obsidian: Approaches to calibration and the analysis of small samples. In *Handheld XRF for Art and Archaeology*; Shugar, A.N., Mass, J.L., Eds.; Leuven University Press: Leuven, Belgium, 2012; pp. 401–422.
23. Frahm, E. Validity of "off-the-shelf" handheld portable XRF for sourcing Near Eastern obsidian chip debris. *J. Archaeol. Sci.* **2013**, *40*, 1080–1092. [CrossRef]
24. Frahm, E.; Doonan, R. The technological versus methodological revolution of portable XRF in archaeology. *J. Archaeol. Sci.* **2013**, *40*, 1425–1434. [CrossRef]
25. Milic, M. PXRF characterisation of obsidian from central Anatolia, the Aegean and central Europe. *J. Archaeol. Sci.* **2014**, *41*, 285–296. [CrossRef]
26. Tykot, R.H. Portable X-ray fluorescence spectrometry (pXRF). In *The SAS Encyclopedia of Archaeological Sciences*; López Varela, S.L., Ed.; John Wiley & Sons, Inc.: Hoboken, NJ, USA, 2018; pp. 1–5.
27. Tykot, R.H. Using non-destructive portable X-ray fluorescence spectrometers on stone, ceramics, metals, and other materials in museums: Advantages and limitations. *Appl. Spectrosc.* **2016**, *70*, 42–56. [CrossRef]
28. Tykot, R.H. A decade of portable (hand-held) X-ray fluorescence spectrometer analysis of obsidian in the Mediterranean: Many advantages and few limitations. *Mrs. Adv.* **2017**, *2*, 1769–1784. [CrossRef]
29. Tykot, R.H. Investigating ancient "bronzes:" Non-destructive analysis of copper-based alloys. In *Artistry in Bronze: The Greeks and Their Legacy. XIXth International Congress on Ancient Bronzes*; Daehner, J.M., Lapatin, K., Spinelli, A., Eds.; The J. Paul Getty Museum and the Getty Conservation Institute: Los Angeles, CA, USA, 2017; pp. 289–299.
30. Speakman, R.J.; Shackley, M.S. Silo science and portable XRF in archaeology: A response to Frahm. *J. Archaeol. Sci.* **2013**, *40*, 1435–1443. [CrossRef]
31. Guthrie, J.M.; Ferguson, J.R. XRF Technical Overview. University of Missouri Research Reactor Archaeometry Laboratory, 2013. Available online: https://archaeometry.missouri.edu/xrf_technical.html (accessed on 1 August 2021).

32. Lugliè, C.; Le Bourdonnec, F.-X.; Poupeau, G.; Bohn, M.; Meloni, S.; Oddone, M.; Tanda, G. A map of the Monte Arci (Sardinia island, Western Mediterranean) obsidian primary to secondary sources. Implications for neolithic provenance studies. *Comptes Rendus Palevol* **2006**, *5*, 995–1003. [CrossRef]
33. Tykot, R.H. The sources and distribution of Sardinian obsidian. In *Sardinia in the Mediterranean: A Footprint in the Sea. Studies in Sardinian Archaeology Presented to Miriam S. Balmuth*; Tykot, R.H., Andrews, T.K., Eds.; Monographs in Mediterranean Archaeology; Sheffield Academic Press: Sheffield, UK, 1992; Volume 3, pp. 57–70.
34. Tykot, R.H. Chemical fingerprinting and source-tracing of obsidian: The central Mediterranean trade in black gold. *Acc. Chem. Res.* **2002**, *35*, 618–627. [CrossRef] [PubMed]
35. Tykot, R.H. Mediterranean Islands and Multiple Flows: The Sources and Exploitation of Sardinian Obsidian. In *Method and Theory in Archaeological Obsidian Studies*; Shackley, M.S., Ed.; Advances in Archaeological and Museum Science; Plenum: New York, NY, USA, 1998; Volume 3, pp. 67–82.
36. Tykot, R.H.; Setzer, T.; Glascock, M.D.; Speakman, R.J. Identification and Characterization of the Obsidian Sources on the Island of Palmarola, Italy. *Geoarchaeol. Bioarch. Stud.* **2005**, *3*, 107–111.
37. Tykot, R.H. Geological sources of obsidian on Lipari and artifact production and distribution in the Neolithic and Bronze Age central Mediterranean. *Open Archaeol.* **2019**, *5*, 83–105. [CrossRef]
38. Tykot, R.H.; Iovino, M.R.; Martinelli, M.C.; Beyer, L. Ossidiana da Lipari: Le fonti, la distribuzione, la tipologia e le tracce d'usura. In Proceedings of the Atti del XXXIX Riunione Scientifica dell'Istituto Italiano di Preistoria e Protostoria: Materie prime e scambi nella preistoria italiana, Firenze, Italy, 25–27 November 2004; pp. 592–597.
39. Francaviglia, V. Ancient obsidian sources on Pantelleria (Italy). *J. Archaeol. Sci.* **1988**, *15*, 109–122. [CrossRef]
40. Tufano, E.; D'Amora, A.; Trifuoggi, M.; Tusa, S. L'ossidiana di Pantelleria: Studio di caratterizzazione e provenienza alla luce della scoperta di nuovi giacimenti. In *Atti Della XXXIX Riunione Scientifica dell'Istituto Italiano di Preistoria e Protostoria*; Istituto Italiano di Preistoria e Protostoria: Firenze, Italy, 2006; pp. 391–402.
41. Tufano, E.; D'Amora, A.; Trifuoggi, M.; Tusa, S. L'ossidiana di Pantelleria: Studio di caratterizzazione e provenienza alla luce della scoperta di nuovi giacimenti. In *Atti Della XLI Riunione Scientifica dell'Istituto Italiano di Preistoria e Protostoria*; Istituto Italiano di Preistoria e Protostoria: Firenze, Italy, 2012; pp. 840–849.
42. Tykot, R.H. Geochemical analysis of obsidian and the reconstruction of trade mechanisms in the Early Neolithic period of the western Mediterranean. In *Archaeological Chemistry. Materials, Methods, and Meaning*; Jakes, K., Ed.; ACS Symposium Series; American Chemical Society: Washington, DC, USA, 2002; pp. 169–184.
43. Cattani, M.; Tosi, M.; Tusa, S. La carta archeologica di Pantelleria. Sperimentazione di metodo e nuove prospettive sull'evoluzione della complessità sociale e politica nelle isole del Mediterraneo centrale. In Proceedings of the Seminario di Scienze Antropologiche, Atti di I Convegno sulla Preistoria e Protostoria siciliana, Corleone, Italy, 17–20 July 1997; pp. 121–133.
44. Renfrew, C.; Wagstaff, M. (Eds.) *An Island Polity: The Archaeology of Exploitation in Melos*; Cambridge University Press: Cambridge, UK, 1982.
45. Birò, T.K. More on the state of art of Hungarian obsidians. *Archeometr. Muh.* **2018**, *XV*, 213–224.
46. Rosania, C.N.; Boulander, M.T.; Biro, K.T.; Ryzhov, S.; Trnka, G.; Glascock, M.D. Revisiting Carpathian obsidian. *Antiquity* **2008**, *82*, 318.
47. Tykot, R.H. Prehistoric obsidian travel to and through central Italy. In *Obsidian and the Sea: Evidence, Concepts and Social Implications of Its Maritime Transportation*; Reepmeyer, C., Moutsiou, T., Eds.; Cambridge University Press: Cambridge, UK, 2021; in press.
48. Tykot, R.H. Obsidian procurement and distribution in the central and western Mediterranean. *J. Mediterr. Archaeol.* **1996**, *9*, 39–82. [CrossRef]
49. Tykot, R.H. New approaches to the characterization and interpretation of obsidian from the Mediterranean island sources. In *Materials Issues in Art and Archaeology VI*; Vandiver, P.B., Goodway, M., Druzik, J.R., Mass, J.L., Eds.; Materials Research Society: Warrendale, PA, USA, 2002; Volume 712, pp. 143–157.
50. Tykot, R.H. Sourcing of Sardinian obsidian collections in the Museo Preistorico-Etnografico 'Luigi Pigorini' using non-destructive portable XRF. In *L'ossidiana del Monte Arci nel Mediterraneo. Nuovi apporti sulla diffusione, sui sistemi di produzione e sulla loro cronologia. Atti del 5° Convegno internazionale (Pau, Italia, 27–29 Giugno 2008)*; Lugliè, C., Ed.; NUR: Ales, Italy, 2010; pp. 85–97.
51. Foresta Martin, F.; Tykot, R.H. Characterization and provenance of archaeological obsidian from Pirozza-Spalmatore, a site of Neolithic colonization on the island of Ustica (Sicily). *Open Archaeol.* **2019**, *5*, 4–17.
52. Tykot, R.H. Obsidian finds on the fringes of the central Mediterranean: Exotic or eccentric exchange? In *Exotica in the Prehistoric Mediterranean*; Vianello, A., Ed.; Oxbow Books: Oxford, UK, 2011; pp. 33–44.
53. Tykot, R.H. Obsidian use and trade in the Adriatic. In *The Adriatic, a Sea Without Borders: Communication Routes of Populations in 6000 BC*; Visentini, P., Podrug, E., Eds.; Civici Musei di Udine, Museo Friulano di Storia Naturale: Udine, Italy, 2014; pp. 171–181, 224–225.
54. Tykot, R.H.; Freund, K.P.; Vianello, A. Source analysis of prehistoric obsidian artifacts in Sicily (Italy) using pXRF. In *Archaeological Chemistry VIII*; Armitage, R.A., Burton, J.H., Eds.; American Chemical Society: Washington, DC, USA, 2013; pp. 195–210.
55. Ammerman, A.J.; Polglase, C. New evidence on the exchange of obsidian in Italy. In *Trade and Exchange in Prehistoric Europe*; Oxbow Monograph; Scarre, C., Healy, F., Eds.; Oxbow Books: Oxford, UK, 1993; Volume 33, pp. 101–108.
56. Tykot, R.H.; Ammerman, A.J. New directions in central Mediterranean obsidian studies. *Antiquity* **1997**, *71*, 1000–1006. [CrossRef]

57. Freund, K.P.; Batist, Z. Sardinian obsidian circulation and early maritime navigation in the Neolithic as shown through social network analysis. *J. Isl. Coast. Archaeol.* **2014**, *9*, 364–380. [CrossRef]
58. Tykot, R.H. Islands in the stream: Stone age cultural dynamics in Sardinia and Corsica. In *Social Dynamics of the Prehistoric Central Mediterranean*; Tykot, R.H., Morter, J., Robb, J.E., Eds.; Accordia Specialist Studies on the Mediterranean; Accordia Research Institute, University of London: London, UK, 1999; Volume 3, pp. 67–82.
59. Tykot, R.H. Obsidian artifacts: Origin of the raw material. In *Special Place, Interesting Times: The Island of Palagruža and Transitional Periods in Adriatic Prehistory*; Forenbaher, S., Ed.; Archaeopress: Oxford, UK, 2018; pp. 84–87.
60. Brown, K.A.; Tykot, R.H.; Muntoni, I.M. Obsidian in the Tavoliere, southeastern Italy—A first regional study. *J. Archaeol. Sci. Rep.* **2018**, *20*, 284–292.
61. Ruka, R.; Galaty, M.; Riebe, D.; Tykot, R.H.; Gjipali, I.; Kourtessi-Philippakis, G. pXRF analysis of obsidian artifacts from Albania: Crossroads or cul-de-sac? *J. Archaeol. Sci. Rep.* **2019**, *24*, 39–49. [CrossRef]
62. Tykot, R.H. Appendix I: Obsidian provenance. In *Ustica I: The Results of the Excavations of the Regione Siciliana Soprintendenza ai Beni Culturali ed Ambientali Provincia di Palermo in Collaboration with Brown University in 1990 and 1991*; Holloway, R.R., Lukesh, S.S., Eds.; Archaeologia Transatlantica; Brown University: Providence, RI, USA, 1995; Volume XIV, pp. 87–90.
63. Foresta Martin, F.; Di Piazza, A.; D'Oriano, C.; Carapezza, M.L.; Paonita, A.; Rotolo, S.G.; Sagnotti, L. New insights into the provenance of the obsidian fragments of the island of Ustica (Palermo, Sicily). *Archaeometry* **2017**, *59*, 435–454. [CrossRef]
64. Speciale, C.; Iovino, M.R.; Freund, K.; de Vita, S.; Larosa, N.; Battaglia, G.; Tykot, R.H.; Vassallo, S. Obsidian from the site of Piano dei Cardoni, Ustica (Palermo, Italy): New Insights on the Prehistoric Occupation of the Island. *Open Archaeol.* **2021**, *7*, 273–290. [CrossRef]
65. Tykot, R.H.; Foresta Martin, F. Analysis by pXRF of prehistoric obsidian artifacts from several sites on Ustica (Italy): Long-distance open-water distribution from multiple island sources during the Neolithic and Bronze Ages. *Open Archaeol.* **2020**, *6*, 348–392. [CrossRef]
66. Tykot, R.H. Obsidian lithics tested by XRF. In *Tas-Silġ, Marsaxlokk (Malta) I: Archaeological Excavations Conducted by the University of Malta, 1996–2005*; Ancient Near Eastern Studies; Bonanno, A., Vella, N.C., Eds.; Peeters: Leuven, Belgium, 2015; Supplement 48; p. 471.
67. Trump, D. *Skorba and the Prehistory of Malta*; Oxford University Press: Oxford, UK, 1966.
68. Malone, C.; Stoddart, S.; Bonanno, A.; Trump, D. (Eds.) *Mortuary Customs in Prehistoric Malta: Excavations at the Brochtorff Circle at Xaghra (1987–94)*; McDonald Institute Monographs: Cambridge, UK, 2009.

Article

Thermoluminescence Analysis of the Clay Core of Bronze Statues: A Re-Appraisal of the Case Studies of Lupa Capitolina and Other Masterpieces in Rome

Marco Martini * and Anna Galli

Dipartimento di Scienza dei Materiali, Università degli Studi di Milano Bicocca, Via Cozzi 55, 20125 Milano, Italy; anna.galli@unimib.it
* Correspondence: m.martini@unimib.it; Tel.: +39-02-64485166

Abstract: In this work, we present some new results in applying thermoluminescence (TL) dating to the clay core of bronze statues. This is very important, due to the impossibility of directly dating a metal. Very few cases of indirect dating of clay cores by TL are reported in the literature. We re-considered three cases of dating of clay core from important bronzes in Rome. The parameters to be considered were not easy to calculate in the case of the *Lupa Capitolina*. However, its traditionally reported Etruscan origin is definitely ruled out, even if the accuracy in the dating is too low to precisely propose a date of the casting. The comparison with radiocarbon results shows good agreement for a Medieval dating. Two other bronze statues were analysed in order to date their casting by TL; a horse from Musei Capitolini resulted to have been cast in the Greek classical period, excluding its casting in the Rome imperial period. A third study shows that, in particularly favourable situations, TL dating of clay core can give rather precise results. This is the case where in the clay core are present materials that behave like good dosimeters, as generally happens in dating ceramics. Furthermore, the possibility of measuring all the parameters influencing the calculation of the dose rate is essential; both the external radiation sources and the radiation reduction by the water content must be taken into account. This was the case of Saint Peter in the Vatican that turned out to be a cast from the beginning of the XIV century.

Keywords: thermoluminescence; dating; clay-core; bronze statue

Citation: Martini, M.; Galli, A. Thermoluminescence Analysis of the Clay Core of Bronze Statues: A Re-Appraisal of the Case Studies of Lupa Capitolina and Other Masterpieces in Rome. *Appl. Sci.* **2021**, *11*, 7820. https://doi.org/10.3390/app11177820

Academic Editor: Dimitris Mossialos

Received: 31 July 2021
Accepted: 24 August 2021
Published: 25 August 2021

Publisher's Note: MDPI stays neutral with regard to jurisdictional claims in published maps and institutional affiliations.

Copyright: © 2021 by the authors. Licensee MDPI, Basel, Switzerland. This article is an open access article distributed under the terms and conditions of the Creative Commons Attribution (CC BY) license (https://creativecommons.org/licenses/by/4.0/).

1. Introduction

The use of delayed luminescence in dating ceramic objects dates back to the Sixties of the past century. It is based on dosimetric principles; in practice, the first type of delayed luminescence, thermoluminescence (TL), uses the light emitted obtained by heating a sample as a measurement of the previously absorbed radiation dose by the sample under study, since its last heating to high temperature. In the next section, a few details are given to better describe the phenomenon.

The linear relationship between emitted light and accumulated dose is fundamental in making TL a very good dosimetric technique. It is currently used to measure the exposure to radiation of professionals working in radiation fields, such as radiologists. Of course, in these cases the materials used are tailored to present an intense emission as a consequence to radiation exposure, well suited to be detected, as it is detailed in the following.

On the contrary, in the application of TL to dating, the material to be used cannot be chosen. In many ceramics and bricks, quartz is contained in relatively large amounts and acts as a good natural dosimeter [1]. Other minerals, such as feldspars, could be good dosimeters, even if some problems are often present, such as the lack of stability of the source of TL signal in time, the so-called "anomalous fading" [2].

In order to date a ceramic by TL, the total absorbed dose since the last heating that generally corresponds to its making in a kiln can be determined by comparing the light emitted due to the exposure to the natural radiation environment in the centuries to artificial irradiations in laboratory. It must be noted that different radiation doses can be accumulated in different samples of the same age, depending on the dose rate, that is, the higher the dose rate, the more intense the natural irradiation. Therefore, the intensity of the mentioned natural radiation environment must be measured. This can generally be achieved by determining the amount of radioactivity of the sample itself and of the surrounding environment. These measurements and the connected experimental troubles are briefly described in the next sections.

In the last decades, many applications of TL dating have demonstrated their feasibility in determining with acceptable precision the sequence in an archaeological stratigraphy or in determining the various phases of construction, modification and restoration in a historical building.

The basic idea of dating by TL can be applied, in principle, to the material remaining in the interior of a bronze statue after its casting, the so-called clay core. In favourable cases, this material behaves like a ceramic and the procedures used for dating ceramics can be also applied to clay cores. This is extremely important, considering that, with very few exceptions that are not treated here, metal objects cannot be dated by absolute techniques.

In this work, the application of TL dating to clay cores is introduced and the specific difficulties deriving from the characteristics of this material, together with the complex determination of the radiation environment, are commented on. Very few cases of application of TL to the dating of clay cores are present in the literature [3,4], mainly due to the many sources of uncertainty mentioned above and to the difficulties in correctly managing the experimental data. The first application of TL techniques to clay cores dates back to 1974, when D.W. Zimmermann [3] succeeded in testing the authenticity of core materials from a Bronze Horse of the New York Metropolitan Museum of Art. Other examples of TL applied to clay core are reported, still in the field of authenticity test [5].

In the last decades a few interesting studies of TL dating were carried out in our laboratory (*Lupa Capitolina*, the horse bronze statue from Musei Capitolini and the Saint Peter statue); they are re-considered in this work aiming at obtaining more accurate results by applying new statistical approaches.

The cases of clay-core dating by TL presented here regard important bronze statues in Rome, including the *Lupa Capitolina*. Sometimes, a rather precise age determination was reached; in some others, it was possible to solve the doubt between two different proposed ages. In this paper, a few results already preliminarily presented are discussed, together with new data and some statistical new calculations aimed at better understanding the dates by TL, when compared with dating results obtained by other techniques.

2. Luminescence Dosimetry

As already mentioned, it is possible to record the exposure to a radiation field by the use of delayed luminescence. Two main types of this kind of feature are currently exploited, thermally- and optically-stimulated luminescence, TSL and OSL, respectively.

Thermoluminescence, or thermally-stimulated luminescence (TL or TSL) is a particular way in which a material emits light, i.e., a luminescence phenomenon [6]. It is a long-lived phosphorescence, in that the light emission is retarded due to the presence of metastable levels, which act as charge "traps", where generally these charges are electrons, of the material. In simple terms, the absorption of energy, mainly from ionising radiation, causes the excitation of electrons in the material, some of which are trapped at the previously mentioned metastable levels, the traps. A subsequent heating of the material, de-trapping the electrons, allows the recombination at luminescence centres, with a consequent light emission, the TL. Similarly, OSL is based on the light emission under optical stimulation; of course, the stimulating light must be different, i.e., with a different wavelength, from the detected emitted light.

In this paper, we focus on TL, but most of the presented procedures are very similar in TL and OSL studies. It must be noted, however, that, in case of TL, the measured radiation amount refers to the period since the last high temperature heating, while in OSL the measured radiation refers to the period since the last light exposure. It can then be applied to materials that have remained in the dark since then, such as sediments. More details on OSL fundamentals and applications can be found elsewhere [6].

When it was discovered that the amount of TL was somehow proportional to the previously absorbed radiation dose, the possibility of using TL as a dosimetric technique was manifested. It was only in 1960 that the practical procedure for quantitative TL measurement was exploited and many TL materials have since then been studied and developed to be used in radiation dosimetry. The result of a TL measurement is the so-called "glow curve" (see Figure 1), which reports the intensity of the emitted light as a function of the temperature; the presence of a peak is related to a previous charge trapping at a certain site. It is strongly dependent on the heating rate, as the position of a peak moves toward higher temperature with increased heating rate.

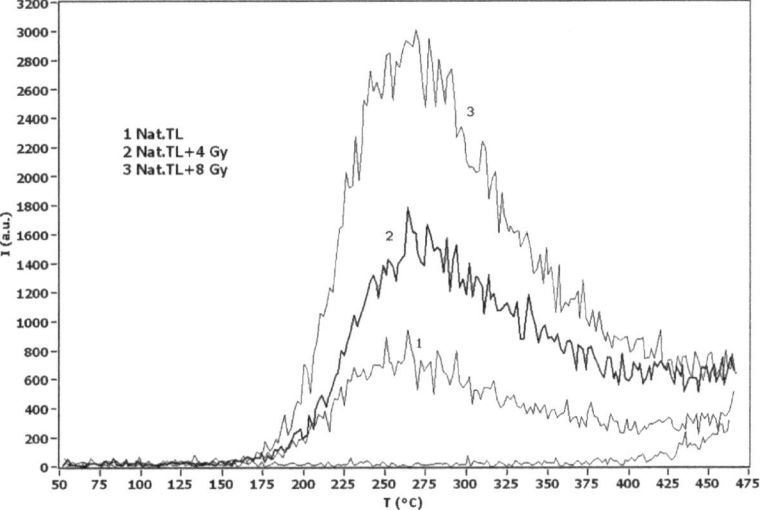

Figure 1. TL glow-curves, example of a clay core extracted from the *Lupa Capitolina*.

The main characteristics needed by a TL material are: (i) the linearity between emitted TL and absorbed dose, or at least a good knowledge of the relationship between TL and absorbed dose, (ii) the presence of peaks in the glow curve at suitable temperature and (iii) the emission wavelength in the visible or near UV region.

3. Thermoluminescence Dating

A particular kind of TL dosimetry is TL dating that was developed in the 1960s. It rapidly became one of the most diffused dating techniques, somehow complementary to radiocarbon dating; TL is used to date inorganic materials, mainly ceramics, while radiocarbon can be applied to organic materials. Luminescence dating has also turned out to be useful in different fields apart from archaeology and historical architecture, in particular in accident dosimetry, while OSL is widely exploited in sediment dating.

In both TL and OSL dating, the materials whose luminescence can be measured are, of course, the minerals naturally present in the objects to be dated; quartz and feldspars are the most diffused minerals contained in ceramics. Looking at TL, their properties are generally good enough, even if the emitted TL is not always proportional to the absorbed dose, due to possible variations in the TL efficiency that must be checked. Besides, in feldspars, the already mentioned phenomenon of "anomalous fading" is often present; it consists in the de-trapping of electrons not due to thermal stimulation and it results in the erasure of part of the TL signal before its measurements. The ways of dealing with such complex behaviours, in order to correctly determine the absorbed dose, have been deeply treated [2].

Looking at the natural radiation field, it comes from both outside, the cosmic rays, and from inside the Earth, the natural radionuclides. These latter are mainly Potassium 40, Uranium 238 and Thorium 232, the last two together with their decay products that are also radioactive, so that they constitute two radioactive families, originating a number of alpha, beta and gamma emissions.

Due to its decay, the intensity of a radioactive substance, its "activity", measured in decays per second, varies with time. In the case of natural radionuclides, being their decay time in the order of 109 years, the activity can be considered constant for relatively "short" times, as they are when treating historic periods (less than 104 years). The interaction of ionising radiation with matter results in a transfer of energy that depends on the type of emitted particles and on the absorbing medium.

The possibility of dating pottery and other materials using TL is based on the same principles on which dosimetry is ground. In fact, the quartz and feldspars crystals usually present in clays act as TL dosimeters; when the clay is heated, to make a pottery or a brick, all the traps are emptied. From that moment, a new filling of the traps starts due to the irradiation by the natural radioactive elements contained in the pottery and in the surrounding environment.

To summarize these mechanisms, there is a fundamental TL dating equation:

$$\text{Age} = \frac{\text{Palaeodose}}{\text{Dose} - \text{rate}} \tag{1}$$

The palaeodose is the total absorbed dose since the last heating at high temperature that generally coincides with the making of the pottery (it can sometimes be due to accidental heating, such as in the case of a fire). It is calculated from a comparison between the "natural" TL produced by the irradiation by the natural radioactivity and the "artificial" TL due to laboratory irradiation with artificial sources, whose intensity is known (see Figure 1).

The dose rate is due to internal radioactive impurities in the object to be dated and external impurities in the surrounding environment, e.g., the burial soil in an archaeological excavation, or the building itself when dating bricks. The internal dose is given by alpha and beta radiation, while the external dose comes from the environmental gamma radiation and, to a lesser extent, to cosmic rays; more details can be found in the specialized literature on the topic [2]. In most cases, the internal radioactivity gives 70–80% of the total dose rate. This constitutes a sizable advantage, because uncertainties in the past environment give limited errors in the calculation of the age. As we explain here, this is not always true when dating clay cores, where it can be a relevant source of uncertainty.

With potteries of Roman–Greek periods, we are dealing with a palaeodose ranging from a few grays to a few tens of grays, where the gray (Gy) is the unit for absorbed dose, i.e., energy per unit mass, and corresponds to 1 Joule per kilogram of absorbing material. The dose rate is usually within the range from 1 milligray per year (mGy/a) to 10 milligrays per year (mGy/a).

As regards the evaluation of the dose rate, due to the very low levels of natural radioactivity, very sensitive techniques have been developed. It must also be taken into account that the effective dose rate is dependent on the water contents of the sample and the soil; this is because one can calculate the exposure to radiation when the radioactive contents of the sample and the environment are known. However, the water present in the pores of the sample absorbs part of the energy emitted by the radionuclides; this means that the dose effectively absorbed by the sample is reduced due to the presence of water. As a result, it is necessary to know, with the highest possible accuracy, the amount of water present in the sample in the centuries, which cannot always be achieved. Typically, there may have been changes in the position of the sample to be dated and in the water content in the past. This source of uncertainty can be a limiting factor when dealing with clay-core dating.

4. TL Dating of Clay Cores: Materials and Method

In the preceding section, the main steps of TL dating and the sources of uncertainty are shortly summarized. It is important to consider how these factors can limit the application of TL dating to clay cores. The precise evaluation of both palaeodose and dose rate requires considering the various factors affecting the calculations.

As regards the measurement of the palaeodose, it must be remembered that the composition of the various clay cores can widely vary from case to case. In fact, the materials that remain in a bronze statue after its casting by the lost-wax technique can be composed not only by clay, but also by many materials of various origin; they were added to allow the attenuation of the huge variations in volume, as an effect of high temperature changes during the casting and after it. Not only inorganic materials, such as clay, but also organic materials, for example, straw, were put inside the object to be cast, in order to attenuate expansion and contraction deriving from the temperature variations. The effects on the TL properties of the materials that can possibly be extracted from the inside of a bronze statue can be important; the presence of organic material is generally a source of the so-called "spurious TL", that is, a light emission independent of the absorbed dose. This phenomenon can also be present in potteries. It is generally attenuated by carrying out the measurements in an inert atmosphere, typically in nitrogen. This is also conducted with clay cores, but the low intensity of the "good" TL signal, as compared with the spurious TL, can constitute a limiting factor in dating clay cores by TL.

A further important point in dating clay cores by TL, especially meaningful with statues found in archaeological excavations, is the presence of materials that were not submitted to the original casting, typically some soil remains that obviously were not submitted to high temperature heating.

In the cases presented in this work, the material extracted from the clay core was submitted to grain selection in order to apply the "fine grain" technique [2]. This implies that the contribution of the natural alpha dose had to be taken into account; alpha irradiation by an ^{241}Am source was carried out in laboratory in order to determine the efficiency of alpha irradiation with respect to the gamma and beta ones [2].

As concerns the measurement of the dose rate, as we explore in some of the following case studies, it would be very important to know, with a good accuracy, the history of the statue to be dated, in order to estimate the external contribution to the dose rate, that is known not to depend on the clay core, but rather on the environmental radiation field.

A second factor affecting the dose rate is the knowledge of the water content of the clay core in the past, whose effect is the attenuation of the dose rate, hence the increase in the date obtained for the measured clay core, as calculated from the radioactivity content of the clay core itself and its environment.

In the cases presented in this work, the radioactivity content of the clay cores was calculated by alpha counting using ZnS (Ag) detectors [2] and through the measurement of the total Potassium content by flame photometry; the contribution due to the radioactive ^{40}K content is easily calculated on the base of its well-known isotopic concentration [2]. The environmental dose rate, as it is detailed in the following, could not be measured due to the lack of information on the external irradiation in the centuries and only speculations could be proposed.

We summarise the specific application of TL dating to clay cores and its difference when compared with the application to potteries and to bricks as follows:

- the materials that constitute the clay core can contain organic materials and spurious TL is expected, also considering a weak signal often found in clay-core materials;
- the clay core can be contaminated by soil particles, not erased by the casting and, consequently, the measured absorbed dose can be overestimated;

- an accurate calculation of the external dose rate sometimes can hardly be determined, due to the ignorance of the history of the object under study, as opposed to an archaeological excavation, or a brick from a building that has been exposed to the same dose rate for centuries.

Similar speculations are valid for the role of the absorbed water, whose effect is the increase in the calculated date, when known. As discussed above, it is generally possible to estimate, sometimes with good accuracy, the water content for an object coming from an archaeological excavation and, even better, from a building. Very difficult is the estimation of the water content inside a bronze statue in the centuries.

5. TL Dating of Clay Cores: A Reappraisal of Some Case Studies

As mentioned in the introduction, few TL studies of the clay cores of bronze statues have been devoted to determining their authenticity, rather than calculating an absolute dating. Further attempts devoted to dating soon enlightened a series of difficulties, complications and limiting factors, as mentioned above. However, in some favourable or particularly meaningful cases, very important results have been achieved.

We report here three case studies whose results have been discussed in the past and have been recently re-considered, obtaining information more relevant than that from past-published data. They regard (i) the famous *Lupa Capitolina*, that has been always considered as the symbol of the Rome itself, (ii) a bronze horse in natural dimensions, exposed in the Musei Capitolini, like the *Lupa*, that could be a Greek original or a Roman copy, such as many bronze statues in Rome, and iii) the statue of San Peter in the Vatican, whose chronological assignment has been long under discussion.

5.1. Lupa Capitolina: TL Study of the Clay Core

During the extensive restorations underwent in the period from 1997 to 2000, several samples of clay core were taken from the *Lupa Capitolina* and submitted to TL studies with the aim of dating its casting [4]. This goal could not be directly reached, due to the complex environmental irradiation history. Nevertheless, it was possible to demonstrate that the statue was cast with maximum of probability between the VIII and XIV century AD, so the traditionally reported Etruscan origin of the statue was ruled out. Recently, Calcagnile et al. [7] had the opportunity to apply AMS radiocarbon dating on, overall, 34 organic residues sampled form the casting cores within the statue.

In view of these results, we crosschecked the TL dates with the radiocarbon ones. We considered the combination of the thirty-four 14C ages (see Figure 6 in reference [7]).

For what concerns the TL dating, it was possible to have access to the interior of the statue for clay core sampling, by hand, by dry laser ablation and with an endoscope specially adapted [4]. The cores had a homogeneous calcareous marl-like composition, rich in silicate and carbonate components, to which abundant inorganic temper was added [8]. For the experimental details, see Martini and Sibilia [4].

By summarising the achieved results, the TL emission of the samples was characterised by a good sensitivity to radiation, but sometimes by scarce reproducibility; therefore, errors in palaeodose were quite high ($\pm 10\%$). As for the dose rate evaluation, even if the concentrations of uranium and thorium were quite homogeneous and the potassium oxide contents were more scattered (standard deviation, 26% of the mean value). The possibility of enrichment of potassium due to percolation of water could perhaps account for this phenomenon.

Finally, speculations on the mean water content during time were performed; the saturation water of the samples was evaluated, but this measured value was used with great caution, giving account for possible significant variations. Moreover, as we already mentioned in the previous section, the assessment of the ambient contribution to the annual dose rate is problematic and this case study was no exception; the measured value refers only to the conservation in the museum and does not take into account any correction for the attenuation due to the metal layer.

So, as expected, TL dating results were strongly dependent on the attenuation of the external dose by the bronze thickness and on the water content. In such a situation, it was therefore attempted a probabilistic approach to the dating problem by focusing the attention on the five solid samples taken from the inner part of the Lupa (D1343/LL4, D1278/LS, D1342/M1, D1344/M2 and D1226/V1, see Figure 2). For these samples, dust contamination could be excluded and humidity conditions and external irradiation were reasonably more constant during time.

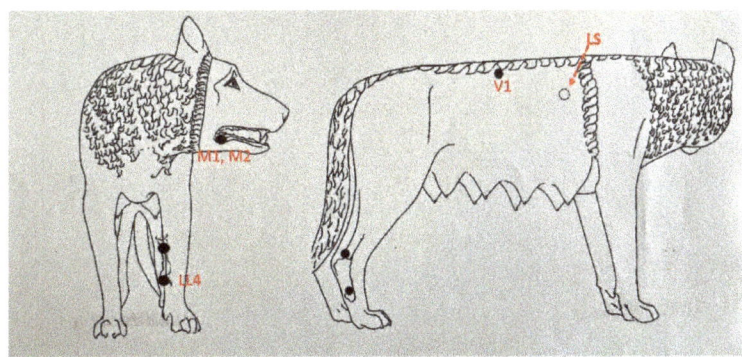

Figure 2. *Lupa Capitolina*, sampling position of the clay core samples.

In Figure 3, the dating results are plotted as a function of the water content, excluding the extreme values relative to dry and saturated conditions, which were very unlikely. The results shown suggest that the casting could not have occurred before the VIII century nor after the XIV century.

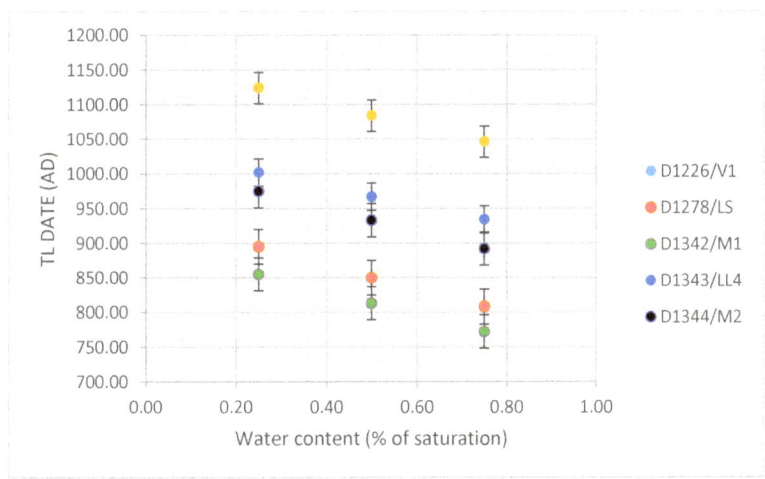

Figure 3. Possible dates for casting (water content, 25–75% of saturation; effective external dose rate, 1.5 ± 0.5 mGy/a).

Putting all the 14C and TL data together (see Figure 4), it appears that, regardless of the chosen percentage of saturation of water content, the TL date of three out of the five samples agree within two standard deviations with the radiocarbon date (namely D1343/LL4, D1344/M2 and D1226/V1).

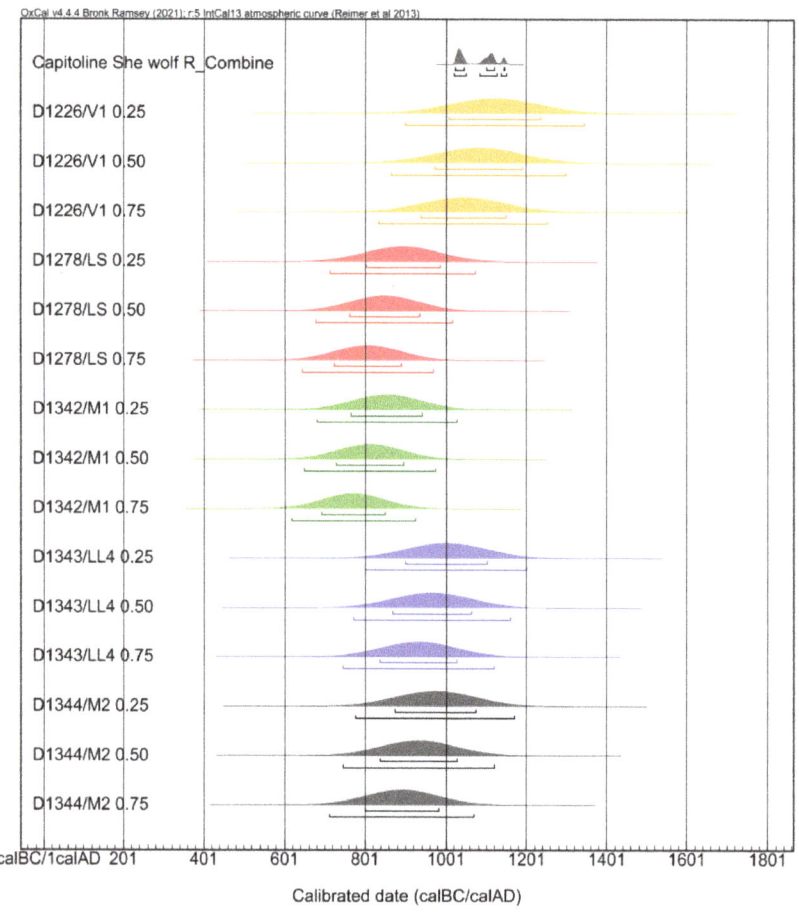

Figure 4. Comparison between the possible TL dates (water content, 25, 50 and 75% of saturation) and the combined radiocarbon date ([7]).

The two samples (D1278/LS and D1342/M1) in disagreement with the combined radiocarbon date are little dependent on the water content. These results seem to put in evidence that if the clay cores are sampled from volumes that have reasonably uniform humidity conditions and external irradiation over time, the estimation of the water content inside the bronze little affects the obtained age.

At this point, we could choose 50% as a value for water content and we applied the R_Combine function of OxCal 4.4.4 [9] to the two sub-set of samples (D1343/LL4, D1344/M2, D1226/V1 and D1278/LS, D1342/M1). The results obtained from the first sub-set (see Figure 5) show an agreement with a confidence level of 95.4% with the radiocarbon age.

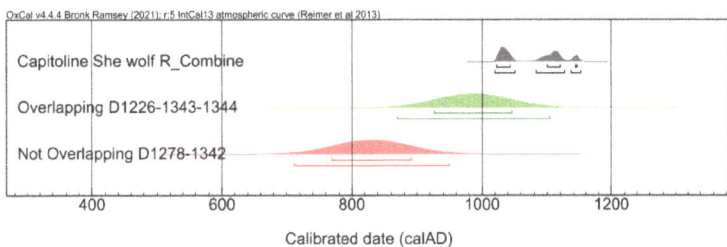

Figure 5. Comparison between the combined radiocarbon date ([7]) and the combination of the two sub-sets of TL dates.

5.2. The Horse Bronze Statue from Musei Capitolini, Rome

In 1894, a number of bronze objects were found in an old cellar in Rome, some of them of great artistic interest. Among them, a beautiful statue of a horse, heavily damaged, was transferred to Musei Capitolini and adequately restored. The statue, see Figure 6, is of natural size and has been exhibited in the Musei since then.

Figure 6. Sculpture, bronze horse, (Capitoline Museum, Rome, Italy). Ph credits Di MM -Opera propria, CC BY-SA 3.0, https://commons.wikimedia.org/w/index.php?curid=22426880, accessed on 31 July 2021.

More than thirty years ago, the statue was submitted to a program of restoration and studies. By the study of the internal clay core, we tried to determine the date of the casting. The problems we faced are presented in Section 4; the expected accuracy could not be very high, either for the material constituent of the clay core, which cannot always be considered a ceramic, or for the difficulties in precisely determining the radiation exposure, due to the uncertainties in the external dose and in the water content. The case of the clay core from the horse statue was rather favourable, at least as regards the type of the constituent material, which did not present any spurious TL, and the calculation of the total absorbed palaeodose could give a rather low uncertainty and similar results among the three samples taken from the interior of the statue. The palaeodose, as measured applying the "fine-grain technique" resulted as $D = 13.2 \pm 0.6$ Gy and was published in a review book [10]. The dating of the statue could not be determined due to the poor information about the dose rate, but a clear indication of a Greek origin was reached, due to the rather elevated value of palaeodose, whatever the dose rate might be, within reasonable values. In the present re-calculation, new information was added and a more reliable assignment of the casting to the Greek period was reached.

We start from the data already obtained in the first measurements, i.e., the internal dose rate, which turned out to be 3.4 ± 0.3 mGy/a. Furthermore, the reported value must be considered as the maximum value of internal dose rate, because any water content would reduce this value. The measurement of the external dose rate, of course, is one of the main sources of uncertainty, together with the water content in the past, as discussed in Section 4. To try to obtain an acceptable value of external dose rate, a dosimetric measurement of the gamma dose rate in the cellar where the statue was found in 1984 was necessary. This value was 4.1 ± 0.3 mGy/a, once more without taking into account the role of water content [4].

To summarise the external dose rate, the statue was exposed to three different dose rates: the first for an unknown period, t_1, until it was abandoned in the cellar where it was found, a second for the period t_2 in the cellar and a known third period, t_3, in Musei Capitolini.

Therefore, having evaluated the dose rates related to the three periods and the palaeodose, it was possible to calculate the statue age by solving a simple parametric equation:

$$\begin{cases} T = t_1 + t_2 + t_3 = \frac{D_1}{d_1} + \frac{D_2}{d_2} + t_3 \\ D = D_1 + D_2 + D_3 \end{cases} \quad (2)$$

where T is the age of statue (the dose rates d_1, d_2 and the period t_3 were evaluated) and D is the palaeodose (D_3 was calculated).

By solving (2), it is possible to obtain the age of casting of the statue as a function of the unknown period t_1:

$$T = mt_1 + q \quad (3)$$

where

$$\begin{cases} m = 1 - \frac{d_1}{d_2} \\ q = \frac{D - D_3}{d_2} + t_3 \end{cases} \quad (4)$$

Figure 7 shows the straight lines obtained in the complete absence of water in the clay cores and with water concentrations varying from 2% to 10%.

Larger percentages are considered unlikely due to the compactness of the material. The intercept of the lines at $t_1 = 0$ provides the age of the horse in the various humidity hypotheses and is the minimum possible in the various cases. The straight line corresponding to 0% water content represents an unrealistic situation but identifies the lower limit of T for the different values of t_1. Reasonable values of water content (5–10%) and outdoor periods ($100 < t_1 < 300$ years) lead to the more likely age between about 2300 and 2400 years and, so, assign the casting of the bronze horse to the classical Greek era.

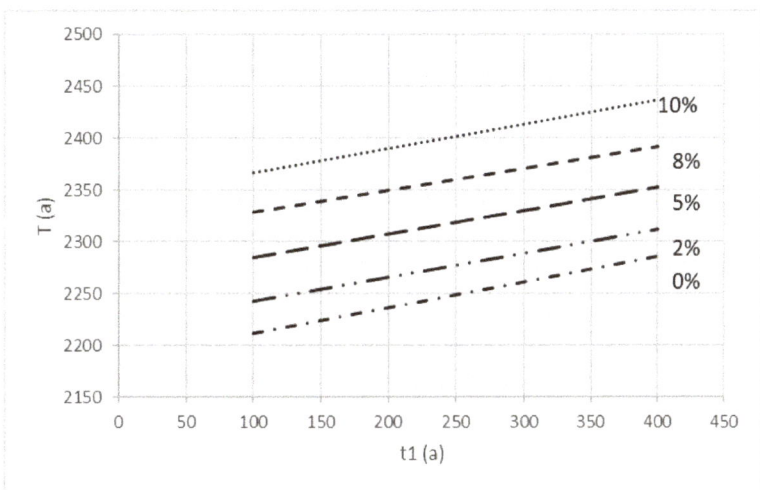

Figure 7. Time elapsed since the casting vs. the period of the outdoor location. Different concentrations of water in the burial environment were considered.

5.3. The Statue of Saint Peter in the Vatican

A case in which a clay core behaves almost like a ceramic was found in the analysis of the statue of Saint Peter in the Vatican Basilica in Rome (see Figure 8).

Figure 8. Bronze Statue of Saint Peter (Arnolfo Di Cambio, St. Peter's Basilica, Rome, Italy). Ph credits Jordiferrer, CC BY-SA 4.0 <https://creativecommons.org/licenses/by-sa/4.0, accessed on 31 July 2021>, via Wikimedia Commons.

The clay core found inside the statue was indeed mostly constituted by clay, which gave a very good dosimetric response [11]. All the three measured samples gave repeatable luminescence intensity and a linear response to the artificial irradiation by the radioactive sources in laboratory. It was then possible to obtain values of the palaeodose in a very narrow Gaussian distribution; the obtained value was 2.7 ± 0.2 Gy. The homogeneity of the material inside the statue allowed a precise determination of the radioactivity content of the clay core; then, the internal dose rate turned out to be 3.3 ± 0.1 mGy/a. The compactness of the clay core allowed to reduce to a very low level the effect of the water content; it must be highlighted that it is almost certain that the statue was put where it is now since its making. As a further consequence, the external dose could be evaluated, taking in due account the attenuation of the bronze itself. All that considered, the annual dose rate resulted to be 3.9 ± 0.2 mGy/a, whose value put in the age equation gave 680 ± 60 years, hence determining a date for the casting of the statue at the beginning of the XIV century. This is in good agreement with the proposal of scholars who attribute it to Arnolfo di Cambio. The dating of this statue is a particularly favourable case of dating clay core taken from the inside of a bronze statue. This is evidently due to the good quality of the clay core itself, in the sense that it was a material very similar to a ceramic, and to the utmost importance of the possibility of determining all the variables contributing to the radiation dose rate, that, in the case of a statue very likely to have been always in the same environment, can all be determined with good precision.

6. Conclusions

Dating clay cores of bronze statues by TL is the only possibility of reaching a chronological determination for the casting of the statues, due to the impossibility of directly dating a metal. Here, it is shown how complex is the application of TL to clay cores. Nonetheless, important results can be achieved by accurate analysis of the experimental data and by a correct statistical management of these data.

The sources of uncertainties can highly reduce the dating precision, either due to the materials, because the clay core is not a ceramic, and due to the difficulties in the calculation of the dose rate. The internal dose rate can be strongly affected by an unknown water content that reduces the dose rate in the centuries, while the external dose rate can be assessed sometimes with large uncertainty, due to the difficult reconstruction of the life of the statue in time. We re-considered three cases of clay-core dating from important bronzes in Rome. The traditionally reported Etruscan origin of *Lupa Capitolina* is definitely ruled out, even if the precision in the dating is too low to allow precise proposals for the date of casting. The statistical comparison with radiocarbon results shows good agreement for a Medieval dating.

A horse from Musei Capitolini is assigned to a casting in the Greek period, rather than a Roman casting.

A third study relative to the statue of Saint Peter in the Vatican shows that, in particularly favourable situations, the dating of clay cores can be similar to the dating of ceramics. It is necessary to find materials in the clay core behaving like good dosimeters and the possibility of measuring all the parameters influencing the calculation of the dose rate is essential, such as in the case of the Saint Peter statue, which has very likely always been kept in the same radiation field.

Simple Summary: The basic idea of dating by TL can be applied, in principle, to the material remaining in the interior of a bronze statue after its casting, the so-called clay core. In favourable cases, this material behaves like a ceramic and the procedures used for dating ceramics can be also applied to clay cores. This is extremely important, considering that, with very few exceptions that are not treated in this paper, metal objects cannot be dated by absolute techniques. In this work, the application of TL dating to clay cores is introduced and the specific difficulties deriving from the characteristics of this material, together with the complex determination of the radiation environment, are commented on.

Author Contributions: Conceptualization, M.M.; writing—original draft preparation, M.M. and A.G.; writing—review and editing, M.M. and A.G.; supervision, M.M. All authors have read and agreed to the published version of the manuscript.

Funding: This research received no external funding.

Institutional Review Board Statement: Not applicable.

Informed Consent Statement: Not applicable.

Acknowledgments: This work was also possible thanks to all researchers that have collaborated with us during the activity of our laboratory. We would like to mention Francesco Maspero particularly, for helping us with the elaboration of data in the case study of *Lupa Capitolina*.

Conflicts of Interest: The authors declare no conflict of interest.

References

1. Preusser, F.; Chithambo, M.L.; Götte, T.; Martini, M.; Ramseyer, K.; Sendezera, E.J.; Susino, G.J.; Wintle, A.G. Quartz as a natural luminescence dosimeter. *Earth-Sci. Rev.* **2009**, *97*, 184–214. [CrossRef]
2. Aitken, M.J. *Thermoluminescence Dating*; Academic Press: London, UK, 1984; pp. 331–351.
3. Zimmermann, D.W.; Yuhas, M.P.; Meyers, P. Thermoluminescence authenticity measurements on core material from the Bronze Horse of the Metropolitan Museum of Art. *Archaeometry* **1974**, *16*, 19–30. [CrossRef]
4. Martini, M.; Sibilia, E. Thermoluminescence Study of the Clay Core of the Lupa Capitolina. *BOREAS* **2009**, *32*, 187–194.
5. Bassett, J. Thermoluminescence dating for European sculpture: A consumer's guide. *Objects Spec. Group Postprints* **2007**, *14*, 32–46.
6. Boetter-Jensen, L.; McKeever, S.W.S.; Wintle, A.G. *Optically Stimulated Luminescence Dosimetry*, 1st ed.; Elsevier: Amsterdam, The Netherlands, 2003; Chapter 5; pp. 119–224.
7. Calcagnile, L.; D'Elia, M.; Maruccio, L.; Braione, E.; Celant, A.; Quarta, G. Solving an historical puzzle: Radiocarbon dating the Capitoline she wolf. *Nucl. Inst. Methods Phys. Res. Sect. B* **2019**, *455*, 209–212. [CrossRef]
8. Lombardi, G. A petrographic study of the casting core of the Lupa Capitolina bronze sculpture (Rome. Italy) and identification of its provenance. *Archaeometry* **2002**, *44*, 601–612. [CrossRef]
9. Ramsey, C.B. Research Lab for Archaeology. Available online: https://c14.arch.ox.ac.uk/oxcal.html (accessed on 2 June 2021).
10. Martini, M.; Sibilia, E.; Spinolo, G. Studio della Termoluminescenza di terra di fusione: Cavallo di bronzo dei Musei Capitolini di Roma. In Proceedings of the 2° Conferenza Internazionale Sulle Prove Non Distruttive, Metodi Microanalitici e Indagini Ambientali Per lo Studio e la Conservazione Delle Opere D'arte di: Istituto Centrale per il Restauro, Perugia, Italy, 17–20 April 1988.
11. Martini, M.; Sibilia, E.; Spinolo, G.; Zelaschi, C. Indirect dating of bronze artifacts using their thermoluminescent clay cores. In Proceedings of the International Symposium "The Ceramics Heritage" of the 8th CIMTEC-World Ceramics Congress and Forum on New Materials, Florence, Italy, 28 June–2 July 1994; pp. 387–391.

Article

Unveiling the Invisible in Uffizi Gallery's Drawing 8P by Leonardo with Non-Invasive Optical Techniques

Alice Dal Fovo, Jana Striova, Enrico Pampaloni and Raffaella Fontana *

National Research Council—National Institute of Optics (CNR-INO), L.go E. Fermi 6, 50125 Firenze, Italy; alice.dalfovo@ino.cnr.it (A.D.F.); jana.striova@ino.cnr.it (J.S.); enrico.pampaloni@ino.cnr.it (E.P.)
* Correspondence: raffaella.fontana@ino.cnr.it

Citation: Dal Fovo, A.; Striova, J.; Pampaloni, E.; Fontana, R. Unveiling the Invisible in Uffizi Gallery's Drawing 8P by Leonardo with Non-Invasive Optical Techniques. *Appl. Sci.* **2021**, *11*, 7995. https://doi.org/10.3390/app11177995

Academic Editor: Asterios Bakolas

Received: 27 July 2021
Accepted: 23 August 2021
Published: 29 August 2021

Publisher's Note: MDPI stays neutral with regard to jurisdictional claims in published maps and institutional affiliations.

Copyright: © 2021 by the authors. Licensee MDPI, Basel, Switzerland. This article is an open access article distributed under the terms and conditions of the Creative Commons Attribution (CC BY) license (https://creativecommons.org/licenses/by/4.0/).

Abstract: Until recently, the study of drawings by old masters has been confined to the art history conservation field. More specifically, scientific investigations of Leonardo's drawings are still very few, possibly due to the latter's extreme fragility and artistic value. However, analytical data are crucial to develop a solid knowledge base of the drawing materials and techniques used by artists in the past. In this work, we report on the application of non-invasive optical techniques on a double-sided drawing by Leonardo belonging to the Uffizi Gallery (8P). We used multispectral reflectography in the visible (Vis) and near-infrared (NIR) regions to obtain a spectral mapping of the drawing materials, to be subsequently integrated with technical information provided by art historians and conservators. Morphological analysis by microprofilometry allowed for the identification of the typical wave-like texture impressed in the paper during the sheet's manufacture, as well as of further paper-impressed traits attributable to the drawing transfer method used by Leonardo. Optical coherence tomography revealed a set of micrometric engraved details in the blank background, which lack any trace of colored material, nor display any apparent relation to the drawn landscape. The disclosure of hidden technical features allowed us to offer new insights into Leonardo's still under-investigated graphic practices.

Keywords: Leonardo da Vinci; drawing; multispectral reflectography; microprofilometry; optical coherence tomography

1. Introduction

Scientific analysis of artworks is often performed within the limits of non-invasiveness requirements, which automatically exclude any material sampling or contact measurement. Meeting such requirements, however, may prove especially challenging when studying paper-based drawings and paintings, whose extremely light-sensitive nature often demands fixed lighting conditions. Minimizing light exposure during measurements without compromising data significance means striking an effective balance between spatial sampling (pixel size) and spectral resolution— an extremely delicate task at best [1–3]. Furthermore, the limited variety of artistic materials typically found in ancient drawings makes it difficult to assess their provenance and authenticity in the absence of proper historical documentation [4]. Numerous studies indicate that analysis of paper-based artworks is best performed when applying a synergic approach that combines non-invasive analytical tools and complementary optical techniques [5]. In the past few decades, spectral imaging processing has been successfully combined with site-specific chemical methods, e.g., Raman spectroscopy [6,7], X-ray fluorescence (XRF) [8,9], fiber optics reflectance spectroscopy (FORS) [10], and particle-induced X-ray emission (PIXE) [11], mainly for the identification and mapping of pigments in medieval illuminated manuscripts and painted books [2,4,8,12,13]. Stratigraphic analysis of miniatures and ancient books has also been performed with infrared thermography (IRT) [14,15] and XRF [15] to highlight the presence of structural defects, such as detachments of the gildings. However, fewer scientific data are available when it comes to drawings on paper, possibly due to their extreme fragility

As a result, analysis of paper-based drawings has thus far been primarily conducted within the historical-conservative field, with only a handful of published studies on drawings by old masters over the 15th–17th centuries contributing to the relevant literature [16–19]. Despite such limitations, the supporting value of scientific data in this regard has proved increasingly crucial in order to gain further insights into drawing materials and techniques used by artists in the past. For instance, art history investigations on Leonardo da Vinci's drawings carried out in the past two decades have successfully integrated analytical measurements to identify constituting materials, delineate the artist's modus operandi, and characterize material deterioration over time [20–22]. Suffice to mention Leonardo's drawings belonging to the Biblioteca Reale in Turin, which were recently analyzed with macro X-ray Fluorescence (MA-XRF), µ-Raman spectroscopy, and atomic force microscopy (AFM) to identify their constituting materials and assess their state of conservation [23,24]. This line of research clearly points to the need for further integration of scientific methods to gain a thorough understanding of Leonardo's creative process.

In this work, we report on the application of non-invasive multi-modal analysis on a double-side drawn sheet by Leonardo, *Drawing 8P* (Figure 1), from the Uffizi Gallery of Florence, Italy. Our study is part of a research campaign carried out by Opificio delle Pietre Dure in collaboration with the National Research Council (CNR) and the National Institute of Nuclear Physics (INFN) within the framework of the exhibition held in Leonardo's birth town (Vinci) to commemorate the five hundredth anniversary of his death (2 May 1519) [25]. The commemoration year also saw other research campaigns, such as the set of investigations on this sheet performed at the University of Bologna (for further reference see [26]). We used multispectral reflectography in the visible (Vis) and near-infrared (NIR) to perform the spectral mapping of the drawing materials, whose elemental composition had been previously characterized by XRF [27]. Morphological micro-features, i.e., not visible to the naked eye, were revealed thanks to the combined application of laser scanning microprofilometry and spectral-domain optical coherence tomography (Sd-OCT). The resulting visible set of hidden details related to the drawing technique allowed us to provide new insights into Leonardo's still under-investigated graphic production.

Figure 1. Leonardo da Vinci, Landscape, 1473, Firenze, Gallerie degli Uffizi, Gabinetto dei Disegni e delle Stampe, inv. 8 P, size 194 × 285 mm, drawing on paper. RGB images of the recto (**a**) and the verso (**b**) acquired via multispectral scanner. The black dashed rectangles highlight the mirrored left-handed (*i*) and the right-handed (*ii*) writings magnified below. Region (*i*) is shown as it is and flipped horizontally (central image) to facilitate the reading of the inscription.

2. Materials and Methods

2.1. Drawing 8P

Drawing 8P (Uffizi Gallery), named after its inventory number, is considered by many to be the first known drawing by the great Tuscan artist and scientist Leonardo da Vinci. He is believed to have depicted the landscape on the front side or recto of the sheet (Figure 1a) when he was twenty-one-years-old and still an apprentice at the prestigious Florentine workshop of Verrocchio. The autograph heading on the top-left margin (Figure 1a, region *i*), "*Dì di s[an]ta Maria della neve/addj 5 dhaghossto1473*", written in Leonardo's typical right-to-left handwriting, indicates the date and the feast associated with the creation of the drawing (5 August 1473, Day of Our Lady of the Snow) [28]. This inscription makes *Drawing 8P* the only known landscape by Leonardo with an autograph date. There is a signature in the lower right corner of the sheet, which is considered apocryphal, possibly added at the end of the 16th century, and related to the collecting history of the drawing. There is no visible sign of watermarks on the paper, thus making it impossible to determine the area of its provenance. Little is known about the drawing's conservation history, due to the absence of attested documentation, such as technical and restoration reports. The circular stamp of the Uffizi collection is still visible in the bottom-left corner of the recto, dating to an old mounting of the drawing on a dark card, which is now missing. The small size of the sheet (around 19 × 28 cm), as well as the complexity of the composition, namely a view from above, suggest that the landscape was made in a workshop rather than outdoors. The scene is constructed through the sum of several elements, starting from the arid terrain with a few isolated trees in the foreground that opens the perspective on a rocky wall with a waterfall (on the right) and a promontory topped by a fortified citadel (on the left). A marshy area with grazing animals on the mainland is represented in the background, with outlines of hills and turreted villages fading into the distance. Leonardo's exceptional drawing skills and speed of execution are made evident by the clear strokes synthesizing each detail, with the sfumato technique, invented by the artist himself, giving the whole scene a sense of vibrant suspension. The back of the sheet or verso (Figure 1b), on the other hand, shows a series of undefined figures and shapes, which have been mostly ignored by scholars in favor of the more evocative landscape on the recto. However, the group of hills with a river and a bridge sketched in the middle of the sheet is considered the initial idea of the landscape later developed on the front [28]. Faded sketches of a female bust and a man's head are also outlined with red chalk on the up-right margin of the folio. Among the traces of geometric figures depicted on the left, one was identified as an early study of Leonardo's famous knots. The left-to-right written inscription on the top margin (Figure 1b, region *ii*) was attributed to Leonardo and thus considered proof of his ability to write with both hands [28].

2.2. Vis-NIR Multispectral Reflectography

Multispectral analysis was performed with the scanner developed at the National Research Council—National Institute of Optics (CNR-INO), allowing for the simultaneous acquisition of 32 narrow-band images (16 VIS + 16 NIR images) through whiskbroom scanning in the range 390–2500 nm [5]. The optical head, i.e., the lighting system and catoptric collecting optics, is placed in a 45°/0° illumination/observation geometry, moving with a step of 250 µm and speed of 500 mm/s. The light reflected from the scanned surface is collected by a square-shaped fiber bundle and delivered to a set of Si and InGaAs photodiodes, each equipped with an individual interferential filter. The autofocus system maintains the optimal target-lens distance during scanning thanks to a high-speed triangulation distance meter and custom-made control software. The instrument's output is a set of perfectly superimposed monochromatic images, which are aberration-free and metrically correct.

The multi-spectral image cube was processed with principal component analysis (PCA) to compress the informative content of large amounts of data in a new, reduced, non-redundant dataset [29]. PCA allows for the expression of the original spectral dataset within a new reference space identified by orthogonal and uncorrelated coordinates, called principal components (PCs), corresponding to linear combinations of the original variables (i.e., the different wavelengths). Since PC variables are hierarchically ordered, the few first PC images are typically representative of the substantial information of the original dataset [1]. Given our research aim, the initial 32 images were reduced to four significant PC images that effectively summarized the salient spectral variations of the drawing. Vis, NIR, and PC images were used to produce false-color (FC) images by combining either a near-infrared, a red, and a green image (NRG-FC), or three PCs (PC-FC). In the first case, traditional NRG→RGB mapping was used (namely, N→R, R→G, and G→B, with "R", "G", and "B" that indicate the red, the green, and the blue channel, respectively, and "N" the near-infrared spectral band). In the second case, three PCs were combined in the trichromatic RGB space. The resulting image provided a detailed outline of the drawings and their similarities—when present, which would otherwise remain undetected during a simple visual inspection, and to a traditional FC.

2.3. Laser Scanning Microprofilometry

Morphological analysis was carried out using a laser scanning micro-profilometer developed at CNR-INO for the survey of a wide range of materials and surfaces. A commercial conoprobe (Conoprobe 1000, Optimet, Jerusalem, Israel), equipped with a 50 mm lens, is moved by a scanning device allowing for measurements on a maximum area of 30 × 30 cm². The profilometer has 1 µm axial resolution, 20 µm lateral resolution, and 8 mm dynamic range. The output is a faithful, high-resolution topographic map of the measured surface, which may be displayed either as a 3D model or as an image. The latter may be further processed through the application of digital filters and rendering techniques to enhance micrometric details and improve their readability.

2.4. Spectral-Domain Optical Coherence Tomography (Sd-OCT)

Cross-sectional analysis was performed with a commercial OCT device, Thorlabs Telesto-II, using a superluminescent diode (central wavelength: 1300 nm, bandwidth: about 100 nm) with axial resolution of 5.5 µm in air, and lateral resolution of 13 µm. The maximum field of view (FOV) is 10.0 × 10.0 mm², with 3.5 mm imaging depth. The detector consists of a spectrograph made of a diffraction grating and a fast camera. The system is controlled via a 64-bit software preinstalled on a high-performance computer. The 3D scanning path probe with integrated video camera performs high-speed imaging (76 kHz) for rapid volume acquisition and live display. The sample stage provides XY translation and rotation of the sample along with axial travel of the probe.

3. Results

3.1. Vis-NIR Multispectral Reflectography

Enhanced instrumental resolution was ensured by raster acquisition performed with oversampling (4x) in both x and y scanning directions, while proper image deconvolution was applied to the datacube. False-color imaging, which was performed on the high-resolution images following the processing described above, revealed traces of the preliminary drawing that outlines the steep cliff on the left of the waterfall and the fortress on the promontory. Details of the rocky wall can be observed in Figure 2, which displays the FC images obtained by combining the NIR -at 1292 nm, the red, and the green images (Figure 2b,d), and PC2, PC3, and PC4 images (Figure 2c,e). According to XRF analysis of the same areas [27], the sketch was drawn with a lead stylus, which is consistent with findings pertaining to other drawings by the same author. The use of a lead stylus is particularly evident in the reported NRG-FC images, where a dark line is visible beneath the red of the main drawing. The same lines appear bright orange in the PC-FC images. Leonardo used to draw with different metal points: lead, which is soft and easy to erase, was used to sketch out drawings that were then refined with a pen; silver and metal alloys were preferred for figure sketches and studies, often completed with lead white highlights [30].

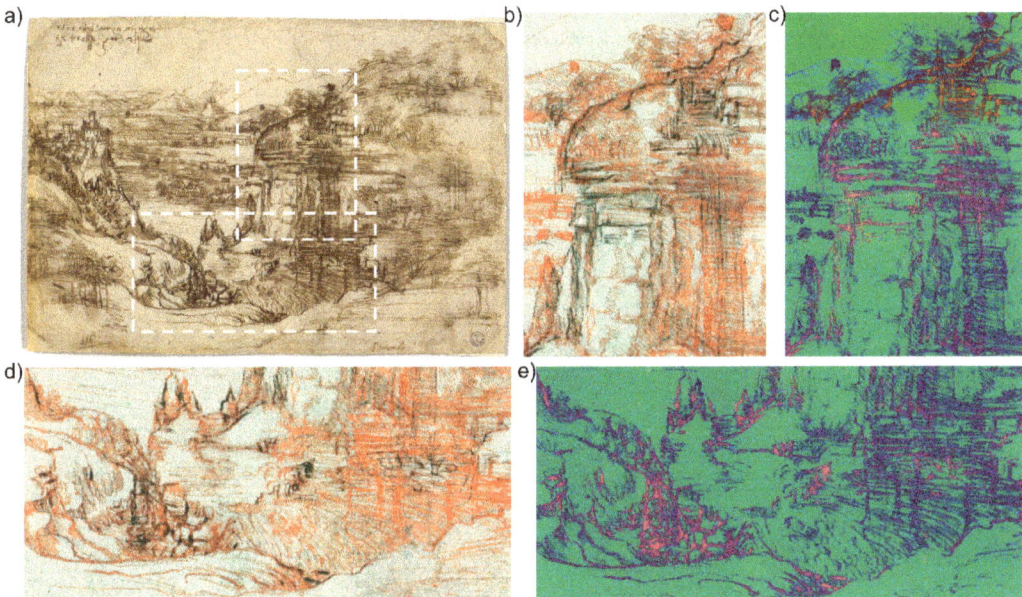

Figure 2. Traditional false-color and trichromatic RGB PC processing reveals the presence of the underdrawing in lead stylus on the recto. (**a**) White dashed rectangles on the RGB image indicate the regions where the preliminary sketch was observed. NRG-FC at 1292 nm (**b**,**d**) and PC-FC (**c**,**e**) images offer a clearer outline of the sketches made with the metal point.

The preliminary sketch in metal point was completed with another tool, possibly a dip pen, in two successive steps [31]. Multispectral analyses showed the presence of two distinct drawings, one outlining the main composition with pale brown lines and the other defining more intense-colored details. The different spectral behavior of the two materials is especially evident in the NIR image at 1600 nm (Figure 3a), where all the main elements of the landscape become transparent except for the outline of the rocks, the fine strokes defining the vegetation in the foreground, and the maelstrom at the bottom of the waterfall. XRF identified common chemical elements in the two types of ink distribution, namely iron, copper, and sulphur. This is consistent with iron-gall ink, and, more specifically, with hydrated green and blue vitriol ($FeSO_4 \cdot nH_2O$ and $CuSO_4 \cdot nH_2O$, respectively), which were typically added to tannic acid solutions to obtain ink [27,32]. The different absorption properties of the two drawings suggest that Leonardo may have used a more diluted, and therefore IR transparent, ink for the main composition, and a more concentrated one for defining specific details [27]. Reflectance spectra (Figure 3d) acquired in the three points shown in Figure 3c confirm the similarities between the two ink compositions. The marked rising in reflectance beyond the red region (inflection point occurring at ca. 730 nm), which accounts for the brownish tone perceived in the visible, is compatible with metal-gall inks [33]. In the analyzed drawing, the overall degree of reflectance appears lower as the ink's hue becomes darker, with a rise occurring in the NIR. Conversely, the lower hiding power of the more diluted iron-gall ink mixture inevitably determines a higher contribution of the paper support, which, in its turn, affects the position of the inflection point in the resulting reflectance spectrum [34]. While there are studies relating the ink's hue to its metal content [35], such a hypothesis requires further verification, especially when considering that artists would often add organic carbon to the mixture to enhance its dark tone. Moreover, recipes for iron-gall inks are numerous and varied, and may include very diverse components and impurities, each undergoing various degradation processes resulting in color changes. In our case, browning and darkening may have resulted from oxidation of ink components into quinonoid structures, as well as degradation products of the Arabic gum in the ink and cellulose in the paper support [32].

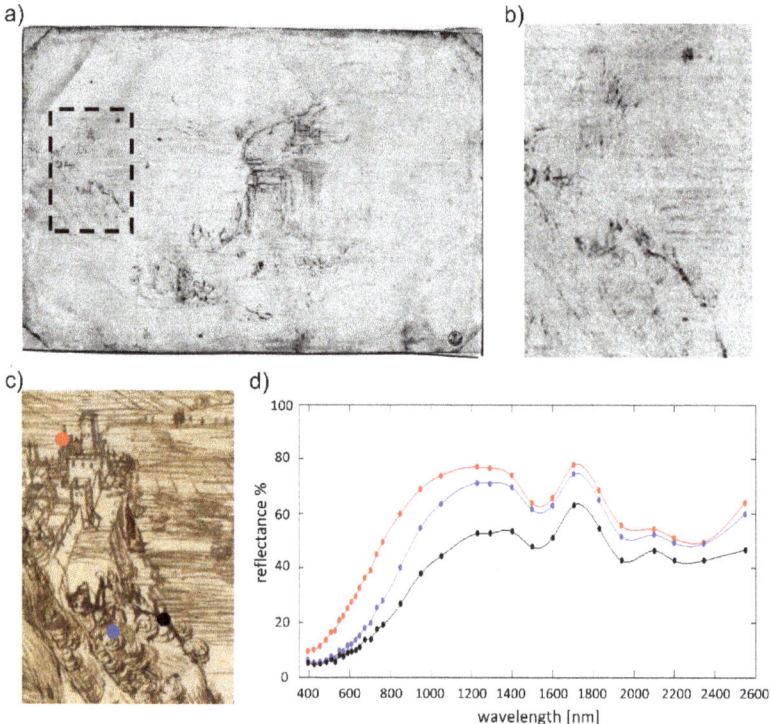

Figure 3. (a) NIR image at 1600 nm showing the absorbing properties of the sketch drawn with one of the two inks; magnified detail of the fortress in the NIR (b) and RGB (c) images; (d) reflectance spectra of the ink in three different points indicated by the red, blue, and black circles in (c).

The use of a similar drawing technique was identified on the verso of the sheet (Figure 4). The main sketch, namely the stream with a bridge and a barely traced outline of hills in the distance, is executed with a dry-point technique [31]. The FC image at 950 nm (Figure 4a) allows for a distinction between this drawing (greenish in appearance) and a few darker details redefined with ink. The latter becomes increasingly transparent in the NIR spectral range and disappears completely at 1600 nm (Figure 4b), consistent with what is observed on the recto, whereas the rocks and the flowing stream in the foreground remain clearly visible. The material used for the main composition, appearing grey in the visible, was identified as lampblack, a dark material typically obtained from the combustion of oils and candles [31]. The soft and grainy appearance of the drawing lines suggests the use of the pastel technique, which originated in France and is believed to have been introduced in Italy by Leonardo himself [36]. The two studies of figures in the upper part of the sheet, both disappearing at 1600 nm, are pen-drawn with brown ink, with indistinct traces of an under-drawing, possibly made with hematite-based red chalk [27], a technique traditionally known as sanguine. These sketches, as well as the bust of a draped figure drawn with the same material, are particularly evident in the detail of the FC-PC image in Figure 4c (light blue traits and yellow traits, respectively).

Figure 4. Multispectral analysis of the verso: (**a**) NRG-FC image at 950 nm evidencing the lampblack pastel drawing (greenish to the eye) and the dark details redefined with iron-gall ink (white dashed rectangles indicating the regions magnified in (**c**,**d**); (**b**) NIR image at 1600 nm; (**c**,**d**) Details of the FC-PC image (PC1-3) showing the sketches of male figures in brown ink appearing light blue, and the bust of a female figure in hematite appearing yellow (**c**); geometric studies in lead point, appearing bluish (**d**).

Fine traces of geometric studies (Figure 4d), executed with a very fine lead point [31], can be seen in the details of the FC-PC image. The superimposed sequence of drawing lines made visible by multispectral imaging suggests that the geometric patterns were the first to be drawn and were then covered by the black pastel landscape and the other ink elements at a later point in time.

3.2. Laser Scanning Microprofilometry

Morphological analysis of the paper surface allowed for the identification of the typical wavy texture impressed in the sheet during the papermaking process. Papermaking between the 15th and the 16th centuries in Europe involved the use of stamping mills consisting of rows of wooden pestles or mallets, which were caused to rise and fall by means of a series of cams to reduce the linen or, more rarely, cotton fabrics to pulp [37,38]. The pulp was suspended in a vat of water. A papermaking mold, i.e., a wooden framework on which a screen made of wires was either placed or strung, was dipped into the vat and scooped up the pulp, thereby trapping the latter within the fine porous screen of the mold. The wires acted as a sieve, filtering out the pulp as the water drained through [39], causing the typical impressions of the wires running sideways ("laid lines") and from top to bottom ("chain lines") on the final sheet. Then, the sodden sheet was transferred onto a wool felt. Alternating wool felts and freshly formed sheets are built up to form a "post", which was eventually transferred to a screw press to remove the excess water, thus impressing the wavy surface texture of the wool felt into the paper. The paper sheets so obtained were finally separated and taken to a loft to dry. The topographic map of the recto (Figure 5a,c), acquired with a sampling step of 50 μm, highlights the presence of seven chain lines with an interval of 3.5–3.7 cm among each, as well as a dense sequence of fine laid lines running parallel to the longer side of the sheet. Interestingly, the NIR reflectance images revealed the presence of seven lines drawn in the same position of the chain lines, which were interpreted as a preliminary grid (by way of example, see NIR image at 1292 nm in Figure 5b,d) [30]. The material used to trace the grid absorbs the Vis-NIR radiation in a manner similar to the above-mentioned preparatory drawing, corroborating the lead point hypothesis [18]. Another series of spaced lines etched parallel to the long side of the sheet appears to have been imprinted freehand by Leonardo with what seems to have been a blind stylus, considering both the width and irregularity of the strokes, and the absence of drawing traces.

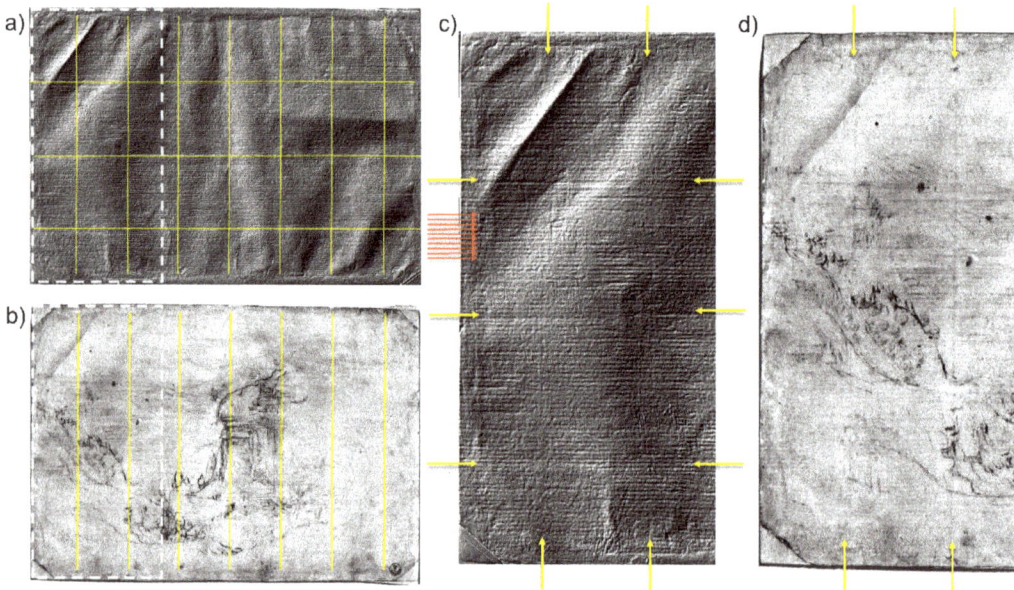

Figure 5. Visualization of the preparatory grid (yellow lines) in the micrometric topographic map (19 × 28 cm) rendered as an image (**a**) and in the NIR image at 1292 nm (**b**). The magnification of the region (19 × 9 cm) highlighted by the white dashed rectangle allows for a clearer identification of the drawn grid lines (pointed out by the yellow arrows in (**c**,**d**)), as well as the fine horizontal laid lines (red arrows in (**c**)) impressed in the paper by the felt.

The 3D model also highlighted the presence of deep incisions at the bottom of the sheet near the profile of the raised terrain in the foreground (Figure 6). These traces may be a by-product of the transferring method used, namely the interposition of transparent paper to transfer the preparatory drawing or assemble sketches to compose a whole scene [40]. In his *Trattato della Pittura*, Leonardo often reports using transparent paper, or even flat glass, not only to copy the preparatory drawing but also to verify the correspondence between the final work and the model previously copied by direct observation [41]. This particular method is still under-researched, due to both the almost total loss of the original materials and the analytic difficulty of detecting its traces. In fact, scholars have often interpreted drawings transferred by means of transparent paper as indirect incisions resulting from cartoon transposition, or even carbon-copying [30].

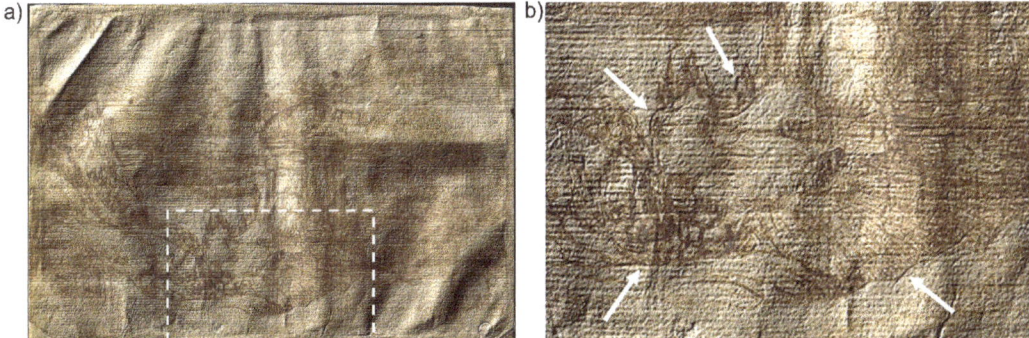

Figure 6. Topographic map of the drawing superimposed on the recto RGB image acquired with the multispectral scanner. The area highlighted by the dashed rectangle (7 × 11 cm) (**a**) is magnified in (**b**); the white arrows indicate the incisions profiling the main elements in the foreground, attributed to the method of transferring the sketch with the use of transparent paper.

3.3. Spectral-Domain Optical Coherence Tomography (Sd-OCT)

OCT tomocubes (5 × 5 × 0.45 mm^3, voxel size 3.5 µm^3) allowed for the areal and cross-sectional visualization of micrometric features related to the artistic technique possibly used by Leonardo. Of particular interest are the traits engraved in the white background with no trace of colored material and no relation to the drawn landscape (Figure 7a,b). Some of these lines are interpreted as superficial scratches ascribed to mechanical damage [31], while others, likely drawn with a blind stylus, seem to outline graphic forms, such as a triangle and an indefinite sketch. The latter's OCT tomocube (Figure 7b) shows a circular concavity, 1.5 mm in diameter and 40–45 µm-deep, which may indicate the use of a pointed tool in this area. Another possible explanation is that they were impressed in the paper while they were being drawn on another superimposed sheet, which would account for their hardly understandable location in the sky above the landscape. Further investigation and comparison with other drawings may clarify this point.

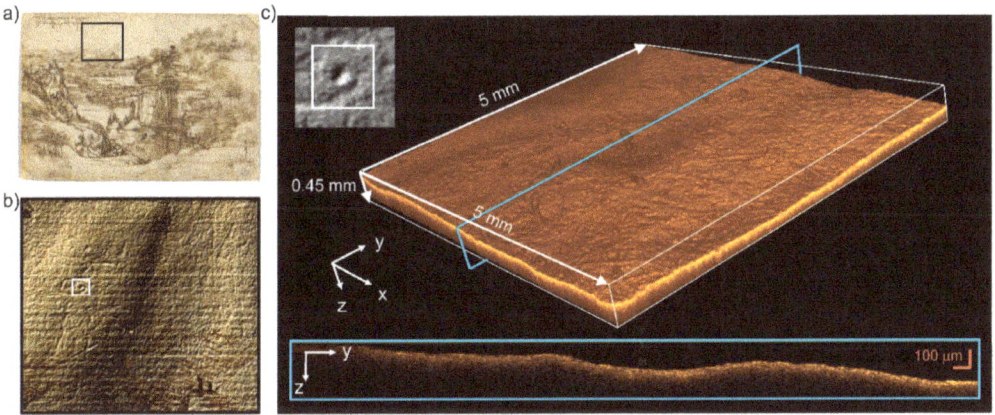

Figure 7. OCT analysis of the blind traits in the background. (**a**) RGB image detail showing the region of interest (ROI, black rectangle); (**b**) topographic map superimposed on the RGB image showing the magnified ROI (the white square showing the area measured with OCT); (**c**) tomocube of the circular impression—light blue rectangle highlighting the position of the z-y section reported below.

4. Discussion and Conclusions

At present there are very few scientific studies supporting the historical-artistic interpretation of paper-based artefacts, particularly those by Leonardo da Vinci. Our results offer a useful contribution to the literature on the drawing technique used by Leonardo in his early production. *Drawing 8P* was analyzed with complementary optical techniques that fully preserve its material integrity. Scientific data were accurately integrated with technical information provided by art historians and conservators involved in the measurement campaign to obtain an exhaustive and reliable characterization of the examined artwork.

False-color and PCA processing of the multispectral images revealed traces of a preliminary sketch of the landscape on the recto, which was drawn with a lead stylus, consistently with other drawings by the same author. Reflectance imaging and spectroscopy highlighted the presence of two distinct superimposed drawings, both made with iron-gal ink, and yet displaying different absorption properties. This seems to indicate that Leonardo used a diluted ink solution for the main composition, and a more concentrated one for the definition of specific details. Furthermore, our analysis revealed a similar superimposition of materials in the main sketch on the verso, also attributed to Leonardo, who outlined the scene with a lampblack pastel and then redefined it with ink and pen. The identification of the pastel technique in the early work of Leonardo is particularly significant, as it is believed that the artist himself was responsible for the introduction of said technique to Italy. The other sketches, caricatures, and writings observed on the backside of the sheet are made with ink, red chalk (sanguine), lead point, lampblack, and metal stylus. The concurrent presence of such diverse tools and materials suggests that they are the result of occasional drawing exercises not only by Leonardo but also by other artists attending the workshop of Verrocchio in the same period, or even later. The poor definition of the least visible sketches, in particular, does not allow for their precise attribution. The only exception are the sanguine sketches, which can be attributed to a later author, since the material used was extremely rare in the years comprised between the time the landscape was drawn and 1480 [42].

Morphological analysis of the sheet by microprofilometry showed the typical wavy texture and chain line/laid line impressions related to the papermaking method available at that time. The 3D model also highlighted the presence of deep incisions, possibly resulting from the use of transparent paper for transferring the preparatory drawing onto the final sheet—a method described by Leonardo himself, yet still poorly documented in terms of scientific investigations.

OCT analysis allowed for the visualization of micrometric details of traits engraved in the white background with no trace of coloured material and no apparent relation to the drawn landscape.

Our work provides corroborating evidence of the suitability of multimodal application of non-invasive techniques to gain new insights into Leonardo's under-investigated modes of creation. The systematic application of complementary analytical methods proved essential for a deeper understanding of such complex and extremely valuable artworks.

Author Contributions: Conceptualization, R.F., A.D.F. and J.S.; methodology, R.F. and A.D.F.; software, E.P.; validation, R.F., J.S. and E.P.; formal analysis, R.F. and A.D.F.; investigation, R.F. and A.D.F.; resources, R.F.; writing—original draft preparation, A.D.F.; writing—review and editing, A.D.F.; visualization, R.F., J.S. and E.P.; supervision, R.F.; project administration, R.F. All authors have read and agreed to the published version of the manuscript.

Funding: This research was funded by Regione Toscana (POR FSE 2014–2020, "Giovanisì", Intervention Program "CNR4C", CUP B15J19001040004).

Institutional Review Board Statement: Not applicable.

Informed Consent Statement: Not applicable.

Acknowledgments: Marco Ciatti, Cecilia Frosinini, Conservator L. Montalbano from Opificio delle Pietre Dure are gratefully acknowledged.

Conflicts of Interest: The authors declare no conflict of interest.

References

1. Bacci, M.; Casini, A.; Cucci, C.; Muzzi, A.; Porcinai, S. A study on a set of drawings by Parmigianino: Integration of art-historical analysis with imaging spectroscopy. *J. Cult. Herit.* **2005**, *6*, 329–336. [CrossRef]
2. Delaney, J.K.; Ricciardi, P.; Glinsman, L.D.; Facini, M.; Thoury, M.; Palmer, M.; de la Rie, E.R. Use of imaging spectroscopy, fiber optic reflectance spectroscopy, and X-ray fluorescence to map and identify pigments in illuminated manuscripts. *Stud. Conserv.* **2014**, *59*, 91–101. [CrossRef]
3. Thomson, G. The museum environment. In *Butterworth-Heinemann series in Conservation and Museology*, 2nd ed.; Butterworth-Heinemann: Oxford, UK, 1986; pp. 16–45.
4. Ricciardi, P.; Delaney, J.K.; Glinsman, L.; Thoury, M.; Facini, M.; de la Rie, E.R. Use of visible and infrared reflectance and luminescence imaging spectroscopy to study illuminated manuscripts: Pigment identification and visualization of underdrawings. *O3A* **2009**, *7391*, 739106.
5. Striova, J.; Dal Fovo, A.; Fontana, R. Reflectance imaging spectroscopy in heritage science. *La Riv. Nuovo Cim.* **2020**, *43*, 515–566. [CrossRef]
6. Legrand, S.; Ricciardi, P.; Nodari, L.; Janssens, K. Non-invasive analysis of a 15th century illuminated manuscript fragment: Point-based vs. imaging spectroscopy. *Microchem. J.* **2018**, *138*, 162–172. [CrossRef]
7. Lauwers, D.; Cattersel, V.; Vandamme, L.; Van Eester, A.; De Langhe, K.; Moens, L.; Vandenabeele, P. Pigment identification of an illuminated mediaeval manuscript De Civitate Dei by means of a portable Raman Equipment. *J. Raman Spectrosc.* **2014**, *45*, 1266–1271. [CrossRef]
8. De Viguerie, L.; Rochut, S.; Alfeld, M.; Walter, P.; Astier, S.; Gontero, V.; Boulch, F. XRF and reflectance hyperspectral imaging on a 15th century illuminated manuscript: Combining imaging and quantitative analysis to understand the artist's technique. *Herit. Sci.* **2018**, *6*, 1–11. [CrossRef]
9. Van der Snickt, G.; De Nolf, W.; Vekemans, B.; Janssens, K. μ-XRF/μ-RS vs. SR μ-XRD for pigment identification in illuminated manuscripts. *Appl. Phys. A Mater. Sci. Process.* **2008**, *92*, 59–68. [CrossRef]
10. Aceto, M.; Agostino, A.; Fenoglio, G.; Idone, A.; Gulmini, M.; Picollo, M.; Ricciardi, P.; Delaney, J.K. Characterisation of colourants on illuminated manuscripts by portable fibre optic UV-visible-NIR reflectance spectrophotometry. *Anal. Methods* **2014**, *6*, 1488–1500. [CrossRef]
11. Kakuee, O.; Fathollahi, V.; Oliaiy, P.; Lamehi-Rachti, M.; Taheri, R.; Jafarian, H.A. External PIXE analysis of an Iranian 15th century poetry book. *Nucl. Instrum. Methods Phys. Res. B Beam Interact. Mater. At.* **2012**, *273*, 178–181.
12. Delaney, J.K.; Facini, M.; Glinsman, L.D.; Thoury, M. Application of imaging spectroscopy to the study of illuminated manuscripts. In Proceedings of the American Institute for Conservation 37th Annual Meeting, Los Angeles, CA, USA, 19–20 May 2009.
13. Melo, M.J.; Otero, V.; Vitorino, T.; Araújo, R.; Muralha, V.S.; Lemos, A.; Picollo, M. A spectroscopic study of brazilwood paints in medieval books of hours. *Appl. Spectrosc.* **2014**, *68*, 434–443. [CrossRef] [PubMed]
14. Doni, G.; Orazi, N.; Mercuri, F.; Cicero, C.; Zammit, U.; Paoloni, S.; Marinelli, M. Thermographic study of the illuminations of a 15th century antiphonary. *J. Cult. Herit.* **2014**, *15*, 692–697. [CrossRef]
15. Mercuri, F.; Gnoli, R.; Paoloni, S.; Orazi, N.; Zammit, U.; Cicero, C.; Marinelli, M.; Scudieri, F. Hidden text detection by infrared thermography. *Restaurator* **2013**, *34*, 195–211.
16. Milota, P.; Reiche, I.; Duval, A.; Forstner, O.; Guicharnaud, H.; Kutschera, W.; Merchel, S.; Priller, A.; Schreiner, M.; Steier, P.; et al. PIXE measurements of Renaissance silverpoint drawings at VERA. *Nucl. Instrum. Meth. Phys. Res. B* **2008**, *266*, 2279–2285. [CrossRef]
17. Radtke, I.M.; Berger, A.; Görner, W.; Ketelsen, T.; Merchel, S.; Riederer, J.; Riesemeier, H.; Roth, M. Spatially resolved synchrotron-induced X-ray fluorescence analyses of metal point drawings and their mysterious inscriptions. *Spectrochim. Acta B* **2004**, *59*, 1657–1662.
18. Tanimoto, S.; Verri, G. A note on the examination of silverpoint drawings by nearinfrared reflectography. *Stud. Conserv.* **2009**, *54*, 106–116. [CrossRef]
19. Reiche, I.; Radtke, M.; Berger, A.; Görner, W.; Merchel, S.; Riesemeier, H.; Bevers, H. Spatially resolved synchrotron radiation induced X-ray fluorescence analyses of rare Rembrandt silverpoint drawings. *Appl. Phys. A* **2006**, *83*, 169–173. [CrossRef]
20. Bambach, C.C. On the role of scientific evidence in the study of Leonardo' drawings. In *Leonardo Da Vinci's Technical Practice: Paintings, Drawings and Influence*; Menu, M., Ed.; Hermann: Paris, France, 2014; pp. 223–253.
21. Ambers, J.; Higgitt, C.; Saunders, D. *Italian Renaissance Drawings: Technical Examination and Analysis*; London Archtype; British Museum: London, UK, 2010.
22. Frosinini, C.; Montalbano, L.; Piccolo, M. *Leonardo e Raffaello, per Esempio. Disegni e Studi D'artista. Catalogue of the Exhibition Held At Palazzo Medici Riccardi–Florence*; Mandragora: Florence, Italy, 2008.
23. Bicchieri, M.; Biocca, P.; Caliri, C.; Romano, F.P. New discoveries on Leonardo da Vinci drawings. *Microchem. J.* **2020**, *157*, 1–6. [CrossRef]
24. Bicchieri, M.; Biocca, P.; Caliri, C.; Romano, F.P. Complementary MA-XRF and μ-Raman results on two Leonardo da Vinci drawings. *X-ray Spectrum.* **2021**, *50*, 401–409. [CrossRef]

25. Barsanti, R. Introduction. In *Leonardo in Vinci: At the Origins of the Genius, Catalogue of the Exhibition Held at the Museo Leonardiano, Vinci, Italy*; Giunti: Milano, Italy, 2019.
26. Marani, P. *Leonardo, Anatomia dei Disegni. Sistema Museale di Ateneo*; Università di Bologna: Bologna, Italy, 2019.
27. Ruberto, C.; Mandò, P.A.; Taccetti, F. X-ray fluorescence scanning analysis. In *Leonardo in Vinci: At the Origins of the Genius, Catalogue of the Exhibition Held at the Museo Leonardiano, Vinci, Italy*; Barsanti, R., Ed.; Giunti: Milano, Italy, 2019.
28. Barsanti, R. Leonardo's landscape of 1473. Research and interpretations. In *Leonardo in Vinci: At the Origins of the Genius, Catalogue of the Exhibition Held at the Museo Leonardiano, Vinci, Italy*; Barsanti, R., Ed.; Giunti: Milano, Italy, 2019.
29. Geladi, P.; Grahn, H.F. Multivariate image analysis. In *Encyclopedia of Analytical Chemistry: Applications, Theory and Instrumentation*; John Wiley & Sons: Hoboken, NJ, USA, 1996.
30. Frosinini, C. Recto and verso, rightwise and leftwise. Drawing 8P in the Uffizi Gallery: In search of the meaning. In *Leonardo in Vinci: At the Origins of the Genius, Catalogue of the Exhibition Held at the Museo Leonardiano, Vinci, Italy*; Barsanti, R., Ed.; Giunti: Milano, Italy, 2019.
31. Montalbano, L. Inks, metal points, chalks and "pastels". An analysis of Leonardo's drawing from a technical-scientific point of view. In *Leonardo in Vinci: At the Origins of the Genius, Catalogue of the Exhibition Held at the Museo Leonardiano, Vinci, Italy*; Barsanti, R., Ed.; Giunti: Milano, Italy, 2019.
32. Carme Sistach, M.; Gibert, J.M.; Areal, R. Ageing of laboratory irongall inks studied by reflectance spectrometry. *Restaurator* **1999**, *20*, 151–166. [CrossRef]
33. Bruni, S.; Caglio, S.; Guglielmi, V.; Poldi, G. The joined use of n.i. spectroscopic analyses–FTIR, Raman, visible reflectance spectrometry and EDXRF–to study drawings and illuminated manuscripts. *Appl. Phys. A* **2008**, *92*, 103–108. [CrossRef]
34. Aceto, M.; Calà, E. Analytical evidences of the use of iron-gall ink as a pigment on miniature paintings. *Spectrochim. Acta A* **2017**, *187*, 1–8. [CrossRef] [PubMed]
35. Jembrih-Simbürger, D.; Desnica, V.; Schreiner, M.; Thobois, E.; Singer, H.; Bovagnet, K. Micro-XRF analysis of watercolours and ink drawings by Albrecht Dürer in the Albertina in Vienna. *Technè* **2005**, *22*, 32–37.
36. Burns, T. *The Invention of Pastel Painting*; Archetype: London, UK, 2007.
37. Hunter, D. *Papermaking: The History and Technique of an Ancient Craft*; Dover Publications: New York, NY, USA, 1978.
38. Thibault, X.; Bloch, J.F. Structural analysis by X-ray microtomography of a strained nonwoven papermaker felt. *Text. Res. J.* **2002**, *72*, 480–485. [CrossRef]
39. Bloom, J.M. Papermaking: The historical diffusion of an ancient technique. In *Mobilities of Knowledge*; Jöns, H., Meusburger, P., Heffernan, M., Eds.; Springer: Cham, Switzerland, 2017; pp. 51–66.
40. Frosinini, C. Carte lucide nella trattistica d'arte e nelle fonti. In *Carte Lucide e Carte Trasparenti Nella Pratica Artistica tra Otto e Novecento: Uso, Conservazione e Restauro, Proceedings of Convegno Internazionale di Studio, Tortona, Italy, 3–4 October 2014*; Scotti Tosini, A., Ed.; Fondazione Cassa di Risparmio di Tortona: Torona, Italy; Associazione Pellizza da Volpedo ONLUS: Volpedo, Italy; Opificio delle Pietre Dure: Firenze, Italy, 2016; pp. 14–28.
41. Da Vinci, L. *Libro di Pittura*; Pedretti, C., Ed.; Giunti: Firenze, Italy, 1995.
42. Meder, J. *The Master of Drawing, Winslow Ames*; Abaris Books, Inc.: New York, NY, USA, 1978.

MDPI
St. Alban-Anlage 66
4052 Basel
Switzerland
Tel. +41 61 683 77 34
Fax +41 61 302 89 18
www.mdpi.com

Applied Sciences Editorial Office
E-mail: applsci@mdpi.com
www.mdpi.com/journal/applsci

www.ingramcontent.com/pod-product-compliance
Lightning Source LLC
LaVergne TN
LVHW070042120526
838202LV00101B/397